Jeremy Narby grew up in Canada and Switzerland, studied history at the University of Canterbury and received a Ph.D. in anthropology from Stanford University. For two years he lived in the Peruvian Amazon studying the Ashaninca Indians and their methods of using the forest's resources. Since 1989 Narby has worked for Nouvelle Planète. Jeremy Narby lives in Switzerland with his wife and three children.

THE COSMIC SERPENT, DNA AND THE ORIGINS OF KNOWLEDGE

JEREMY NARBY

WEIDENFELD & NICOLSON

A W&N PAPERBACK

First published in 1995 as
Le serpent cosmique, l'ADN et les origines du savoir
by Georg Editeur SA, Geneva

Note on translation: The author wrestled the
text from French into English with the assistance
of Jon Christensen

First published in Great Britain by Victor Gollancz in 1998
This paperback edition published in 1999 by Weidenfeld & Nicolson,
an imprint of Orion Books Ltd,
Carmelite House, 50 Victoria Embankment,
London EC4Y 0DZ

An Hachette UK company

Published by arrangement by Jeremy P. Tarcher, Inc.,
A division of Penguin Putnam, Inc., New York

23 25 27 29 30 28 26 24

A CIP catalogue record for this book
is available from the British Library.

ISBN 978-0-7538-0851-1

Printed and bound in Great Britain by
Clays Ltd, Elcograf S.p.A.

The Orion Publishing Group's policy is to use papers that
are natural, renewable and recyclable products and
made from wood grown in sustainable forests. The logging
and manufacturing processes are expected to conform to
the environmental regulations of the country of origin.

www.orionbooks.co.uk

•

THOSE
WHO
LOVE
WISDOM
MUST
INVESTIGATE
MANY
THINGS.

•

HERACLITUS

To

Rachel, Arthur,
Loïk, and Gaspar

CONTENTS

THE COSMIC SERPENT,
DNA AND THE ORIGINS
OF KNOWLEDGE

Chapter 1

FOREST TELEVISION

The first time an Ashaninca man told me that he had learned the medicinal properties of plants by drinking a hallucinogenic brew, I thought he was joking. We were in the forest squatting next to a bush whose leaves, he claimed, could cure the bite of a deadly snake. "One learns these things by drinking ayahuasca," he said. But he was not smiling.

It was early 1985, in the community of Quirishari in the Peruvian Amazon's Pichis Valley. I was twenty-five years old and starting a two-year period of fieldwork to obtain a doctorate in anthropology from Stanford University. My training had led me to expect that people would tell tall stories. I thought my job as an anthropologist was to discover what they really thought, like some kind of private detective.

During my research on Ashaninca ecology, people in Quirishari regularly mentioned the hallucinatory world of *ayahuasqueros*, or shamans. In conversations about plants, animals, land, or the forest, they would refer to ayahuasqueros as the source of knowledge. Each time, I would ask myself what they really meant when they said this.

I had read and enjoyed several books by Carlos Castaneda on the uses of hallucinogenic plants by a "Yaqui sorcerer." But I knew that the anthropological profession had largely discredited Castaneda, accusing him of implausibility, plagiarism, and fabrication.[1] Though no one explicitly blamed him for getting too close to his subject matter, it seemed clear that a subjective consideration of indigenous hallucinogens could lead to problems within the profession. For me, in 1985, the ayahuasqueros' world represented a gray area that was taboo for the research I was conducting.

Furthermore, my investigation on Ashaninca resource use was not neutral. In the early 1980s, international development agencies were pouring hundreds of millions of dollars into the "development" of the Peruvian Amazon. This consisted of confiscating indigenous territories and turning them over to market-oriented individuals who would then develop the "jungle" by replacing it with cow pastures. Experts justified these colonization and deforestation projects by saying that Indians didn't know how to use their lands rationally.[2] I wanted to argue the contrary by doing an economic, cultural, and political analysis showing the *rational* nature of Ashaninca resource use. To emphasize the hallucinatory origin of Ashaninca ecological knowledge would have been counterproductive to the main argument underlying my research.

After two months in the field, I experienced an unexpected setback. I had left Quirishari for ten days to renew my visa in Lima. On returning to the community I was met with indifference. The following day, during an informal meeting in front of the house I was staying in, people asked whether it was true that I was going to return to my country to become a doctor. The question surprised me, as I usually described my future profession as "anthropologist," rather than "doctor," to avoid any confusion with "medical practitioner." It turned out that several employees of the government's development project, the Pichis-Palcazu Special Project, had come to Quirishari in my absence and inquired about my activities. In answer the people showed them my file containing samples of medicinal plants. The project employees then scolded the inhabitants of Quirishari for being naive Indians—did they not realize that I was going to become a doctor and make a fortune with their plants?

In fact I had been classifying these plants to show that the tropical forest, which seemed "unused" to the experts flying over it in airplanes, represented a pharmacy for the Ashaninca, among other things. I had explained this to the inhabitants of Quirishari at the beginning of my stay. However, I knew that any further explanation would only confirm their suspicions, as I was truly going to become a "doctor." I therefore proposed to put an immediate stop to the collection of medicinal plants and to entrust the contentious file to the community's primary school. This settled the matter and dissipated the tension in the air—but it also removed one of the empirical bases on which I had been hoping to build a thesis demonstrating the rational nature of Ashaninca resource use.

After four months of fieldwork I left Quirishari to visit the neighboring community of Cajonari, a seven-mile walk through the forest. The inhabitants of Cajonari had let it be known that it was not fair for Quirishari to have the exclusive monopoly on the anthropologist who was giving "accounting" classes. These were actually informal arithmetic lessons that I had started to teach at the community's request.

People in Cajonari gave me a warm welcome. We spent several evenings telling stories, singing for my tape recorder, and drinking manioc beer, a milky liquid that tastes like cold, fermented potato soup. During the day we did arithmetic, worked in the gardens, or listened to the songs taped the previous evening. Everyone wanted to listen to their own performance.

One evening in front of a house half a dozen men and I were drinking manioc beer and chatting in the twilight. The conversation veered to the question of "development," a daily subject in the valley since the arrival of the Pichis-Palcazu Special Project and its $86 million budget. In general the Ashaninca expressed frustration, because they were continually being told that they did not know how to produce for a market, whereas their gardens were full of potential products and everyone dreamed of making a little money.

We were discussing the differences between Ashaninca agriculture and "modern" agriculture. I already understood that, despite their apparent disorderliness, indigenous gardens were polycultural masterpieces containing up to seventy different plant species that were mixed chaotically, but never innocently. During the conversation I praised their practices and ended up expressing my astonishment at their botanical mastery, asking, "So how did you learn all this?"

A man named Ruperto Gomez replied, "You know, brother Jeremy, to understand what interests you, you must drink ayahuasca."

I pricked up my ears. I knew that ayahuasca was the main hallucinogen used by the indigenous peoples of Western Amazonia. Ruperto, who was not turning down the calabashes of beer, continued in a confident tone: "Some say it is occult, which is true, but it is not evil. In truth, ayahuasca is the television of the forest. You can see images and learn things." He laughed as he said this, but no one else smiled. He added, "If you like, I can show you sometime."[3]

I replied that I would indeed be interested. Ruperto then launched into a comparison between my "accounting" science and his "occult" science. He had lived with the Shipibo, the northern neighbors reputed for their powerful medicine. He had followed a complete ayahuasquero apprenticeship, spending long months in the forest eating only bananas, manioc, and palm hearts and ingesting huge quantities of hallucinogens under the watchful eye of a Shipibo ayahuasquero. He had just spent eight years away from Cajonari, over the course of which he had also served in the Peruvian army—a source of personal pride.

On my part, I had certain prejudices about shamanism. I imagined the "veritable" shaman to be an old wise person, traditional and detached—somewhat like Don Juan in the Castaneda books. Ruperto the wanderer, who had learned the techniques of another tribe, did not correspond to my expectations. However, no old wise person had stepped up to initiate me, and I was not going to be choosy. Ruperto had made his proposal spontaneously, publicly, and as part of a bargain. In return I was to give him a special "advanced" accounting course. So I accepted his offer, especially since it seemed that it might not materialize once the effects of the beer had worn off.

Two weeks later I was back in Quirishari, when Ruperto appeared for his first private lesson. He told me before leaving, "I will return next Saturday. Prepare yourself the day before, eat neither salt nor fat, just a little boiled or roasted manioc."

He returned on the appointed day with a bottle full of a reddish liquid that was corked with an old corncob. I had not followed his instructions, because, deep down, I did not really take the matter seriously. The idea of not eating certain foods before an event seemed to me a superstition. For lunch I had nibbled a bit of smoked deer meat and some fried manioc.

Two other people had agreed to take ayahuasca under Ruperto's direction. At nightfall, the four of us were sitting on the platform of a quiet house. Ruperto lit a cigarette that he had rolled in notebook paper and said, "This is *toé*." He passed it around. If I had known at that point that toé is a kind of datura, I would perhaps not have inhaled the smoke, because datura plants are powerful and dangerous hallucinogens that are widely recognized for their toxicity.[4] The toé tasted sweet, though the cigarette paper could have been finer.

Then we each swallowed a cup of ayahuasca. It is extremely bitter and tastes like acrid grapefruit juice. Thirty seconds after swallowing it, I felt nauseated.

I did not take notes or keep time during the experience. The description that follows is based on notes taken the next evening.

First Ruperto sprayed us with perfumed water (*agua florida*) and tobacco smoke. Then he sat down and started to whistle a strikingly beautiful melody.

I began seeing kaleidoscopic images behind my closed eyes, but I was not feeling well. Despite Ruperto's melody, I stood up to go outside and vomit. Having disposed of the deer meat and fried manioc remnants, I returned feeling relieved. Ruperto told me that I had probably eliminated the ayahuasca also and that, if I wanted, I could have some more. He checked my pulse and declared me strong enough for a "regular" dose, which I swallowed.

Ruperto started whistling again as I sat down in the darkness of the platform. Images started pouring into my head. In my notes I describe them as "unusual or scary: an agouti [forest rodent] with bared teeth and a bloody mouth; very brilliant, shiny, and multicolored snakes; a policeman giving me problems; my father looking worried. . . ."

Deep hallucinations submerged me. I suddenly found myself

surrounded by two gigantic boa constrictors that seemed fifty feet long. I was terrified. "These enormous snakes are there, my eyes are closed and I see a spectacular world of brilliant lights, and in the middle of these hazy thoughts, the snakes start talking to me without words. They explain that I am just a human being. I feel my mind crack, and in the fissures, I see the bottomless arrogance of my presuppositions. It is profoundly true that I am just a human being, and, most of the time, I have the impression of understanding everything, whereas here I find myself in a more powerful reality that I do not understand at all and that, in my arrogance, I did not even suspect existed. I feel like crying in view of the enormity of these revelations. Then it dawns on me that this self-pity is a part of my arrogance. I feel so ashamed that I no longer dare feel ashamed. Nevertheless, I have to throw up again."

I stood up feeling totally lost, stepped over the fluorescent snakes like a drunken tightrope walker, and, begging their forgiveness, headed toward a tree next to the house.

I relate this experience with words on paper. But at the time, language itself seemed inadequate. I tried to name what I was seeing, but mostly the words would not stick to the images. This was distressing, as if my last link to "reality" had been severed. Reality itself seemed to be no more than a distant and one-dimensional memory. I managed nonetheless to understand my feelings, such as "poor little human being who has lost his language and feels sorry for himself."

I have never felt so completely humble as I did at that moment. Leaning against the tree, I started throwing up again. In Ashaninca, the word for ayahuasca is *kamarampi*, from the verb *kamarank*, "to vomit." I closed my eyes, and all I could see was red. I could see the insides of my body, red. "I regurgitate not a liquid, but colors, electric red, like blood. My throat hurts. I open my eyes and feel presences next to me, a dark one to my left, about a yard away from my head, and a light one to my right, also a yard away. As I am turned to my left, I am not bothered by the dark presence, because I am aware of it. But I jump when I become aware of the light presence to my right, and I turn to look at it. I can't really see it with my eyes; I feel so bad, and control my reason so little, that I do not really want to see it. I remain lucid enough to understand that I am not truly vomiting blood. After a while I start wondering what to do. I have so little control that I abandon myself to the instructions that seem to be coming from outside me: now it is time to stop vomiting, now it is time to spit, to blow nose, to rinse mouth with water, not to drink water. I am thirsty, but my body stops me from drinking."

I looked up and saw an Ashaninca woman dressed in a traditional long cotton robe. She was standing about seven yards away from me, and she seemed to be levitating above the ground. I could see her in the darkness, which had become clear. The quality of the light reminded me of those night scenes in movies which are filmed by day with a dark filter: somehow, not really dark, because glowing. As I looked at this woman, who was staring at me in silent clear darkness, I was once again staggered by this people's familiarity with a reality that turned me upside down and of which I was totally ignorant.

"Still very confused, I reckon I have done everything, including rinse my face, and I feel amazed that I have been able to do all this by myself. I leave the tree, the two presences and the levitating woman, and I return to the group. Ruperto asks, 'Did they tell you not to drink water?' I answer, 'Yes.' 'Are you drunk (mareado)?' 'Yes.' I sit down and he resumes his song. I have never heard more beautiful music, these slender staccatos that are so high-pitched they verge on humming. I follow his song, and take flight. I fly in the air, thousands of feet above the earth, and looking down, I see an all-white planet. Suddenly, the song stops, and I find myself on the ground, thinking: 'He can't stop now.' All I can see are confused images, some of which have an erotic content, like a woman with twenty breasts. He starts singing again, and I see a green leaf, with its veins, then a human hand, with its veins, and so on relentlessly. It is impossible to remember everything."

Gradually, the images faded. I was exhausted. I fell asleep shortly after midnight.

ANTHROPOLOGISTS
AND SHAMANS

The main enigma I encountered during my research on Ashaninca ecology was that these extremely practical and frank people, living almost autonomously in the Amazonian forest, insisted that their extensive botanical knowledge came from plant-induced hallucinations. How could this be true?

The enigma was all the more intriguing because the botanical knowledge of indigenous Amazonians has long astonished scientists. The chemical composition of ayahuasca is a case in point. Amazonian shamans have been preparing ayahuasca for millennia. The brew is a necessary combination of two plants, which must be boiled together for hours. The first contains a hallucinogenic substance, dimethyltryptamine, which also seems to be secreted by the human brain; but this hallucinogen has no effect when swallowed, because a stomach enzyme called monoamine oxidase blocks it. The second plant, however, contains several substances that inactivate this precise stomach enzyme, allowing the hallucinogen to reach the brain. The sophistication of this recipe has prompted Richard Evans Schultes, the most renowned ethnobotanist of the twentieth century, to comment: "One wonders how peoples in primitive societies, with no knowledge of chemistry or physiology, ever hit upon a solution to the activation of an alkaloid by a monoamine oxidase inhibitor. Pure experimentation? Perhaps not. The examples are too numerous and may become even more numerous with future research."[1]

So here are people without electron microscopes who choose, among some 80,000 Amazonian plant species, the leaves of a bush containing a hallucinogenic brain hormone, which they combine with a vine containing substances that inactivate an enzyme of the digestive tract, which would otherwise block the hallucinogenic effect. And they do this to modify their consciousness.

It is as if they knew about the molecular properties of plants *and* the art of combining them, and when one asks them how they know these things, they say their knowledge comes directly from hallucinogenic plants.[2]

Not many anthropologists have looked into this enigma[3]—but the failure of academics to consider this kind of mystery is not limited to the Amazon. Over the course of the twentieth century, anthropologists have examined shamanic practices around the world without fully grasping them.

A brief history of anthropology reveals a blind spot in its studies of shamanism.

IN THE NINETEENTH CENTURY, European thinkers considered that some races were more evolved than others. Charles Darwin, one of the founders of the theory of evolution, wrote in 1871: "With civilised nations, the reduced size of the jaws from lessened use, the habitual play of different muscles serving to express different emotions, and the increased size of the brain from greater intellectual activity, have together produced a considerable effect on their general appearance in comparison with savages."[4]

Anthropology was founded in the second half of the nineteenth century to study "primitive," "Stone Age" societies. Its underlying goal was to understand where "we" Europeans had come from.[5]

The problem for the young discipline was the unreasonable nature of its object of study. According to Edward Tylor, one of the first anthropologists: "Savages are exceedingly ignorant as regards both physical and moral knowledge; want of discipline makes their opinions crude and their action ineffective in a surprising degree; and the tyranny of tradition at every step imposes upon them thoughts and customs which have been inherited from a different stage of culture, and thus have lost a reasonableness which we may often see them to have possessed in their first origin. Judged by our ordinary modern standard of knowledge, which is at any rate a high one as compared to theirs, much of what they believe to be true, must be set down as false."[6]

The question was: How could one study such incoherence scientifically?

The "father of modern anthropology," Bronislaw Malinowski, found the answer by developing a method for the objective analysis of "savages." Called "participant observation," and used to this day, it involves living in close contact with the natives while observing them from a distance. By considering native reality with a *distant gaze,* the anthropologist manages to introduce "law and order into what seemed chaotic and freakish."[7]

From the 1930s onward, anthropology obsessively sought order in its study of others, to elevate itself to the rank of science.[8] In the process it transformed reality into next to incomprehensible discourses.[9]

Here is an extract from Claude Lévi-Strauss's book *The elementary structures of kinship* (1949), one of the texts by which anthropology claimed to attain the rank of science: "For example, in a normal eight-subsection system the grandson would reproduce his father's father's subsection by marriage with the mother's mother's brother's daughter's daughter. The wavering of the Murinbata between the traditional system and the new order ends in practice with the identification of the mother's brother's daughter and the mother's mother's brother's daughter's daughter as the possible marriage partner, i.e., for TJANAMA: *nangala* = *nauola*. Hence, a TJIMIJ man marries a *namij* woman. The father maintains that his daughter is *nalyeri* (which is the 'conventional' subsection). However, a *namij* woman is by kinship *purima*, a 'marriageable' daughter of the sister's son, but according to the subsections she is a 'sister.' Consequently, her daughter is *nabidjin*, for according to the aboriginal rule, formulated in a matrilineal idiom: '*namij* makes *nabidjin*.' From this the conflict arises of whether the subsections are patrilineal or matrilineal."[10]

Just when anthropology thought it had established itself within the scientific community thanks to such "structuralist" discourses, it experienced a fundamental setback. Its object of study, those primitives living outside of time, started to vanish like snow in the sun; by the middle of the twentieth century, it had become increasingly difficult to find "real" natives who had never had any contact with the industrial world. Indeed, such people may never even have existed. As early as the second half of the nineteenth century, the indigenous peoples of the Amazon, for instance, were dragooned on a grand scale into the construction of the industrial world, to which they contributed a vital component, rubber. Since then most of them have used metal tools of industrial origin.

During the 1960s, this crisis plunged anthropology into the doubt and self-criticism of "poststructuralism." Anthropologists came to realize that their presence changed things, that they were themselves sorts of colonial agents, and, worse yet, that their methodology was flawed. Participant observation is a contradiction in terms, because it is impossible to observe people from above while participating in the action at their side, to watch the game from the stands while playing on the field. The anthropological method condemns its practitioners to "dance on the edge of a paradox"[11] and to play the schizophrenic role of the player-commentator. Furthermore, the distant gaze of the anthropologist cannot perceive itself, and those who aspire to objectivity by using it cannot see their own presuppositions. As Pierre Bourdieu put it, objectivism "fails to objectify its objectifying relationship."[12]

Anthropologists discovered that their gaze was a tool of domination and that their discipline was not only a child of colonialism, it also served the colonial cause through its own practice. The "unbiased and supra-cultural language of the observer" was actually a colonial discourse and a form of domination.[13]

The solution for the discipline consisted in accepting that it was not a science, but a *form of interpretation*. Claude Lévi-Strauss himself came to say: "The human sciences are only sciences by way of a self-flattering imposture. They run into an insurmountable limit, because the realities they aspire to understand are of the same order of complexity as the intellectual means they deploy. Therefore they are incapable of mastering their object, and always will be."[14]

ANTHROPOLOGISTS INVENTED the word "shamanism" to classify the least comprehensible practices of "primitive" peoples.

The word "shaman" is originally Siberian. Its etymology is uncertain.[15] In the Tungus language, a *saman* is a person who beats a drum, enters into trance, and cures people. The first Russian observers who related the activities of these saman described them as mentally ill.

From the early twentieth century onward, anthropologists progressively extended the use of this Siberian term and found shamans in Indonesia, Uganda, the Arctic, and Amazonia. Some played drums, others drank plant decoctions and sang; some claimed to cure, others cast spells. They were unanimously considered neurotic, epileptic, psychotic, hysterical, or schizophrenic.[16]

As George Devereux, an authority on the matter, wrote: "In brief, there is no reason and no excuse for not considering the shaman as a severe neurotic and even a psychotic. In addition, shamanism is often also culture dystonic. . . . Briefly stated, we hold that the shaman is mentally deranged. This is also the opinion of Kroeber and Linton."[17]

In the middle of the twentieth century, anthropologists began to realize not only that "primitives" did not exist as such, but that shamans were not crazy. The change came abruptly. In 1949, Claude Lévi-Strauss stated in a key essay that the shaman, far from being mentally ill, was in fact a kind of psychotherapist—the difference being that "the psychoanalyst listens, whereas the shaman speaks." For Lévi-Strauss, the shaman is above all a *creator of order*, who cures people by transforming their "incoherent and arbitrary pains" into "an ordered and intelligible form."[18]

The shaman as creator of order became the creed of a new generation of anthropologists. From 1960 to 1980, the established authorities of the discipline defined the shaman as a creator of order, a master of chaos, or an avoider of disorder.[19]

Of course, things did not happen so simply. Until the late 1960s, several members of the old school continued to claim that shamanism was a form of mental illness,[20] and in the 1970s it became fashionable to present the shaman as a specialist in all kinds of domains who plays "the roles of physician, pharmacologist, psychotherapist, sociologist, philosopher, lawyer, astrologer, and priest."[21] Finally, in the 1980s, a few iconoclasts claimed that shamans were creators of disorder.

So who are these shamans? Schizophrenics or creators of order? Jacks-of-all-trades or creators of disorder?

The answer lies in the mirror. When anthropology was a young science, unsure of its own identity and unaware of the schizophrenic nature of its own methodology, it considered shamans to be mentally ill. When "structuralist" anthropology claimed to have attained the rank of science, and anthropologists busied themselves finding order in order, shamans became creators of order. When the discipline went into a "poststructuralist" identity crisis, unable to decide whether it was a science or a form of interpretation, shamans started exercising all kinds of professions. Finally, some anthropologists began questioning their discipline's obsessive search for order, and they saw shamans as those whose power lies in "insistently questioning and undermining the search for order."[22]

It would seem, then, that the reality hiding behind the concept of "shamanism" reflects the anthropologist's gaze, independently of its angle.

In 1951, around the time Lévi-Strauss was transforming the schizophrenic shaman into the psychoanalyst–creator of order, Mircea Eliade, one of the foremost authorities in the history of religions, published the now classic *Shamanism: Archaic techniques of ecstasy*. To this day, it is the only attempt at a world synthesis on the subject.

Eliade, who was not a trained anthropologist, saw neither

mental illness nor creation of order. Instead he identified astonishing similarities in the practices and concepts of shamans the world over. Wherever these "technicians of ecstasy" operate, they specialize in a trance during which their "soul is believed to leave the body and ascend to the sky or descend to the underworld." They all speak a "secret language" which they learn directly from the spirits, by imitation. They talk of a ladder—or a vine, a rope, a spiral staircase, a twisted rope ladder—that connects heaven and earth and which they use to gain access to the world of spirits. They consider these spirits to have come from the sky and to have created life on earth.[23]

Anthropologists rarely appreciate it when library-based intellectuals use their work without muddying their boots and discover connections that they had not seen. They made no exception with Eliade, rejecting his work because of its "latent mysticism." They accused him of detaching symbols from their contexts, mutilating and distorting the facts, obliterating the sociocultural aspect of the phenomenon and locking it into a mystical dead end. Recently, it was even said that Eliade's notion of celestial flight was "a potentially fascistic portrayal of third world healing."[24]

Nevertheless, despite these criticisms, Eliade understood before many anthropologists that it is useful to take people and their practices seriously and to pay attention to the detail of what they say and do.

SOME ANTHROPOLOGISTS REALIZED that the academic studies of shamanism were going around in circles. This led them to criticize the very notion of "shamanism." Clifford Geertz, for instance, wrote that shamanism is one of those "insipid categories by means of which ethnographers of religion devitalize their data."[25]

However, abandoning the concept of "shamanism," as was done thirty years ago with the notion of "totemism,"[26] will not clarify the reality to which it refers. The difficulty of grasping "shamanism" lies not so much in the concept itself as in the gaze of those who use it. The academic analysis of shamanism will always be the rational study of the nonrational—in other words, a self-contradictory proposition or a cul-de-sac.

Perhaps the most revealing example in this respect is provided by Luis Eduardo Luna, the author of an excellent study of the shamanism of mestizo ayahuasqueros in the Peruvian Amazon, who practice what they call *vegetalismo,* a form of popular medicine based on hallucinogenic plants, singing, and dieting. Luna focuses on the techniques of these shamans and reports their opinions without interpreting them. He writes: "They say that ayahuasca is a doctor. It possesses a strong spirit and it is considered an intelligent being with which it is possible to establish rapport, and from which it is possible to acquire knowledge and power if the diet and other prescriptions are carefully followed." However, Luna writes in a rational language for a rational public ("us"), and it is not rational to claim that certain plants are intelligent beings capable of communication. Luna, who explores the question of "plant-teachers" over several pages, ends up concluding: "Nothing can be said . . . until we have some kind of understanding as to what these people are really talking about, when they say that the plants themselves reveal their properties."[27] One cannot consider that what they say is real, because, in reality as "we" know it, plants do not communicate.

There is the blind spot.

Chapter 3

THE MOTHER OF THE
MOTHER OF TOBACCO
IS A SNAKE

T wo days after my first ayahuasca experience, I was walking in
the forest with Carlos Perez Shuma, my main Ashaninca
consultant. Carlos was forty-five years old and was an experienced
tabaquero-ayahuasquero who had also dealt extensively with missionaries and colonists. We reached a river that we had to cross
and paused. The moment seemed right to ask a few questions,
particularly since Carlos had also participated in the hallucinatory
session two nights previously. "*Tío* [uncle]," I asked, "what are
these enormous snakes one sees when one drinks ayahuasca?"
"Next time, bring your camera and take their picture," he answered. "That way you will be able to analyze them at your
leisure."

I laughed, saying I did not think the visions would appear on
film. "Yes they would," he said, "because their colors are so bright."
With this, he stood up and started wading across the river.

I scampered after him, thinking about what he had just said. It
had never occurred to me that one could seriously consider taking
pictures of hallucinations. I was certain that if I did so, I would
only obtain photos of darkness. But I knew that this would not

prove anything, because he could always question the capacities of my camera. In any case, these people seemed to consider the visions produced by hallucinogenic plants to be at least as real as the ordinary reality we all perceive.

A few weeks later I started recording a series of interviews with Carlos, who had agreed to tell me his life story. The first evening, we sat on the platform of his house, surrounded by the sounds of the forest at night. A kerosene lamp made of a tin can and a cotton wick provided a flickering source of light and gave off blackish fumes.

Despite my training, it was the first time in my life that I interviewed someone. I did not know where to start, so I asked him to start at the beginning.

Carlos was born in the Perene Valley in 1940. He lost his parents when he was five years old in the waves of epidemics that swept the area with the arrival of white settlers. His uncle took care of him for several years. Then he went to an Adventist mission, where he learned to speak, read, and write Spanish.

What follows is an extract from the transcript of this first interview. We talked in Spanish, which is neither his mother tongue nor mine, as a faithful translation reveals:

"My uncle was a tabaquero. *I watched him take lots of tobacco, dry it a bit in the sun, and cook it. I wondered what it could be. 'That's tobacco,' my uncle told me, and once the mixture was good and black, he started tasting it with a little stick. I thought it was sweet, like concentrated cane juice. When he ate his tobacco, he could give people good advice. He could tell them, 'this is good' or 'this is not good.' I don't know what the intellectuals say now, but in those days, all the Adventist missionaries said, 'He is listening to his bats, to his Satan.' He had no*

book to help him see, but what he said was true: 'Everybody has turned away from these things, now they all go to the missionary. I do not know how to read, but I know how to do these things. I know how to take tobacco, and I know all these things.' So when he talked, I listened. He told me: 'Listen nephew, when you are a grown man, find a woman to look after, but before that, you must not only learn how to write, you must also learn these things.'"

"Learn to take tobacco?" I asked.

"Take tobacco and cure. When people would come to him, my uncle would say: 'Why do you ask me to cure you, when you say you know God now that you are at the mission, and I do not know God? Why don't you ask the pastor to pray, since he says he can cure people with prayers? Why don't you go to him?' But he would cure them anyway. He would pull out his coca, start chewing it, and sit down like us here now. Then, he would swallow his tobacco. Meanwhile, I would watch him and ask him what he was doing. The first time I saw him cure, he said: 'Very well, bring me the sick baby.' First, he touched the baby, then took his pulse: 'Ah, I see, he's in a bad way. The illness is here.' Then, he started sucking the spot [suction noise]. Then, he spat it out like this: ptt! Then, again, and a third time: ptt! There, very good. Then he told the mother: 'Something has shocked this little one, so here is a herb to bathe him. After that, let him rest.' The next day, one could already see an improvement in the baby's health. So I took a liking to it and decided to learn. Ooh! The first time I had tobacco, I didn't sleep."

"How old were you?"

"I was eight years old. I thought tobacco was sweet. But it was so bitter that I couldn't even swallow it. My uncle said: 'That's the secret of tobacco.' Then, he showed me everything.

He gave me a tobacco gourd. Little by little, I learned to take it and to resist. Fairly quickly, I stopped vomiting."

"Did your uncle also teach you how to use ayahuasca?"

"No, I learned that later, with my father-in-law. . . ."

Over the following months, I recorded approximately twenty hours on the meanders of Carlos's life. He spoke Spanish better than anybody in Quirishari; in the past, he had taught it to other Ashaninca in an Adventist school. However, his grammar was flexible, and he talked with unexpected rhythms, punctuating his sentences with pauses, gestures, and noises that completed his vocabulary nicely, but that are difficult to put into written English. Furthermore, his narrative style varied from a first-person account to the commentary of a narrator who also plays the roles of the characters. This is no doubt more appropriate for oratory, or radio plays, than for a text.

By taping Carlos's life story, I was not trying to establish the point of view of a "typical" Ashaninca. Rather, I was trying to grasp some specifics of local history by following the personal trajectory of one man. In particular, I was interested in questions of territory in the Pichis Valley: Who owned which lands, and since when? Who used which resources? As it happens, the overall history of the Ashaninca in the twentieth century is closely defined by the progressive expropriation of their territory by outsiders, as Carlos's life story reveals.

Carlos's birthplace, the Perene Valley, was the first Ashaninca region to undergo colonization. By 1940, the majority of indigenous lands in the area had already been confiscated. Ten years later, Carlos the young orphan had followed the mass migration of the Perene Ashaninca toward the Pichis Valley, where the forests were still free of colonists and diseases. After living twenty-six years in this new homeland, Carlos had been elected to the presidency of the congress of the Association of the Indigenous Communities of the Pichis (ACONAP). The goal of this organization was to defend indigenous lands from a new onslaught of colonization. Carlos was forced to abandon his position after four years when he was bitten by a snake. At this point, he retired to Quirishari to cure himself "with ayahuasca and other plants." When I appeared five years later, he was living like a retired politician, satisfied with the tranquillity, but nostalgic for yesteryear's struggles. He did not seem displeased at the idea of confiding his memoirs to a visiting anthropologist.

Over the course of our conversations, I often asked Carlos about the places he had lived, directing the conversation toward the solid ground of social geography. But he would regularly answer in ways that pointed toward shamanism and mythology. For example:

> *"The earthquake in the Perene, was that in 1948 or 1947?"*
> *"1947."*
> *"And were you there at the time?"*
> *"Of course, at that time, I was a young boy. It happened in Pichanaki. It killed three people. Pichanaki was a nice plain, but now there are more than twenty meters of earth burying the old village. It used to be a fertile lowland, good for corn."*
> *"And why was this place called Pichanaki?"*
> *"That's the name that the first natives gave it in the old days, the tabaqueros, the ayahuasqueros. As I have explained to you, it is simply in their visions that they were told that the river is called Pichanaki."*
> *"Ah yes. And 'Pichanaki' means something? All these place names that finish in -aki, like Yurinaki also, what does 'aki' mean?"*

"It means that there are many minerals in the center of these places. The word means 'eye' in our language."

"And 'Picha'?"

"He is called like that, because in the hills, there is a representative of the animals whose name is Picha."

"Ah, 'the eyes of Picha.'"

"Now you see."

I often asked Carlos to explain the origin of place names to me. He would invariably reply that nature itself had communicated them to the ayahuasqueros-tabaqueros in their hallucinations: "That is how nature talks, because in nature, there is God, and God talks to us in our visions. When an ayahuasquero drinks his plant brew, the spirits present themselves to him and explain everything."

Listening to Carlos's stories, I gradually became familiar with some of the characters of Ashaninca mythology. For instance, he often talked of Avíreri: "According to our ancient belief, he is the one of the forest, he is our god. He was the one who had the idea of making people appear." Carlos also referred to invisible beings, called *maninkari,* who are found in animals, plants, mountains, streams, lakes, and certain crystals, and who are sources of knowledge: "The maninkari taught us how to spin and weave cotton, and how to make clothes. Before, our ancestors lived naked in the forest. Who else could have taught us to weave? That is how our intelligence was born, and that is how we natives of the forest know how to weave."

I had not come to Quirishari to study indigenous mythology. I even considered the study of mythology to be a useless and "reactionary" pastime. What counted for me were the hectares confiscated in the name of "development" and the millions of dollars in international funds that financed the operation. With my research, I was trying to demonstrate that true development consisted first in recognizing the territorial rights of indigenous people. My point of view was materialist and political, rather than mystical.[1] So, after nine months in Quirishari, it was almost despite myself that I started reading Gerald Weiss's doctoral dissertation on Ashaninca mythology, entitled *The Cosmology of the Campa Indians of Eastern Peru*—"Campa" being the disparaging word used until recently to designate the Ashaninca, who do not appreciate it.[2]

I discovered as I read this thesis that Carlos was not making up fanciful stories. On the contrary, he was providing me with concise elements of the main cosmological beliefs of his culture, as documented extensively by Weiss in the 1960s.

According to Weiss, the Ashaninca believe in the existence of invisible spirits called maninkari, literally "those who are hidden," who can nonetheless be seen by ingesting tobacco and ayahuasca. They are also called *ashaninka,* "our fellows," as they are considered to be ancestors with whom one has kinship. As these maninkari are also present in plants and animals, the Ashaninca think of themselves as members of the same family as herons, otters, hummingbirds, and so on, who are all *perani ashaninka,* all our fellows long ago.[3]

Some maninkari are more important than others. Weiss distinguishes a hierarchy among these spirits. Avíreri, the god who creates by transformation, is the most powerful of them all. In Ashaninca myths, Avíreri, accompanied by his sister, creates the seasons with the music of his panpipes. He shapes human beings by blowing on earth. Then he wanders with his grandson Kíri, casually transforming human beings into insects, fruit trees, animals, or rock formations. Finally, Avíreri gets drunk at a festival. His malicious sister invites him to dance and pushes him into a hole that she has dug beforehand. Then she pretends to pull him out by throwing him a thread, a cord, and finally a rope, none of which is strong enough. Avíreri decides to escape by digging a tunnel into the underworld. He ends up in a place called "river's end," where a strangler vine wraps itself around him. From there, he continues to sustain his numerous children of the earth. And Weiss concludes: "There *Avíreri* remains to the present day, no more able to move, because of the vine that constrains him."[4]

Finally, Weiss notes in passing: "To be sure, although these accounts are to be classified and referred to as myths, for the Campas they are reliable reports handed down orally from past generations of real happenings, happenings as authentically real as any actual event of past years that someone still remembers or was told about."[5]

I had the same impression as Weiss: My Ashaninca informants discussed mythological characters and events as if they were real. This seemed quite fanciful to me, but I did not say so. As an anthropologist I was trained to respect outlandish beliefs.

THE INHABITANTS of Quirishari had made it clear to me that I was not supposed to gather plant samples. However, I could study their uses of the forest as I pleased, and I could try their plant remedies.

So whenever I had a health problem and people told me they knew of a cure, I tried it. Often the results went beyond not only my expectations, but my very understanding of reality. For instance, I had suffered from chronic back pain since the age of seventeen, having played too much tennis during my adolescence. I had consulted several European doctors, who had used cortisone injections and heat treatment, to no avail. In Quirishari there was a man, Abelardo Shingari, known for his "body medicine." He proposed to cure my back pain by administering a *sanango* tea at the new moon. He warned me that I would feel cold, that my body would seem rubbery for two days, and that I would see some images.

I was skeptical, thinking that if it were really possible to cure chronic back pain with half a cup of vegetal tea, Western medicine would surely know about it. On the other hand, I thought it was worth trying, because it could not be less effective than cortisone injections.

Early one morning, the day after the new moon, I drank the sanango tea. After twenty minutes, a wave of cold submerged me. I felt chilled to the bone. I broke out into a profuse cold sweat and had to wring out my sweatshirt several times. After six rather difficult hours, the cold feeling went away, but I no longer controlled the coordination of my body. I could not walk without falling down. For five minutes I saw an enormous column of multicolored lights across the sky—my only hallucinations. The lack of coordination lasted forty-eight hours. On the morning of the third day, my back pain had disappeared. To this day it has not returned.[6]

I tend not to believe this kind of story unless I have lived it myself, so I am not trying to convince anybody about the effectiveness of sanango. However, from my point of view, Abelardo had pulled off a trick that seemed more biochemical than psychosomatic.

I had several other similar experiences. Each time, I noted that the seemingly fanciful explanations I was given ended up being verified in practice—such as "a tea you drink at the new moon which turns your body to rubber and cures your back pain."

So I began to trust the literal descriptions of my friends in Quirishari even though I did not understand the mechanisms of their knowledge.

Furthermore, by living with them on a daily basis, I was continually struck by their profound practicality. They did not talk of doing things; they did them. One day I was walking in the forest with a man named Rafael. I mentioned that I needed a new handle for my ax. He stopped in his tracks, saying "ah yes," and used his machete to cut a little hardwood tree a few steps off the path. Then he carved an impeccable handle that was to last longer than the ax itself. He spent about twenty minutes doing the bulk of the work right there in the forest and an additional twenty minutes at home doing the adjustments. Perfect work, carried out by eye alone. Up until then, I had always thought that ax handles came from hardware stores.

People in Quirishari taught by example, rather than by explanation. Parents would encourage their children to accompany them in their work. The phrase "leave Daddy alone because he's working" was unknown. People were suspicious of abstract concepts. When an idea seemed really bad, they would say dismissively, *"Es pura teoría"* ["That's pure theory"]. The two key words that cropped up over and over in conversations were *práctica* and *táctica*, "practice" and "tactics"—no doubt because they are requirements for living in the rainforest.

The Ashaninca's passion for practice explains, in part at least, their general fascination for industrial technology. One of their favorite subjects of conversation with me was to ask how I had made the objects I owned: tapes, lighters, rubber boots, Swiss army knife, batteries, etc. When I would reply that I did not know how to make them, nobody seemed to believe me.

After about a year in Quirishari, I had come to see that my

hosts' practical sense was much more reliable in their environment than my academically informed understanding of reality. Their empirical knowledge was undeniable. However, their explanations concerning the origin of their knowledge were unbelievable to me. For instance, on two separate occasions, Carlos and Abelardo showed me a plant that cured the potentially mortal bite of the *jergón* (fer-de-lance) snake. I looked at the plant closely, thinking that it might come in useful at some point. They both pointed out the pair of white hooks resembling snake fangs, so that I would remember it. I asked Carlos how the virtues of the jergón plant had been discovered. "We know this thanks to these hooks, because that is the sign that nature gives."

Once again, I thought that if this were true, Western science would surely know about it; furthermore I could not believe that there was truly a correspondence between a reptile and a bush, as if a common intelligence were lurking behind them both and communicating with visual symbols. To me, it seemed that my "animist" friends were merely interpreting coincidences of the natural order.

ONE DAY at Carlos's house, I witnessed an almost surreal scene. A man called Sabino appeared with a sick baby in his arms and two Peruvian cigarettes in his hand. He asked Carlos to cure the child. Carlos lit one of the cigarettes and drew on it deeply several times. Then he blew smoke on the baby and started sucking at a precise spot on its belly, spitting out what he said was the illness. After about three minutes, he declared the problem solved. Sabino thanked him profusely and departed. Carlos called after him, placing the second cigarette behind his ear: "Come back any time."

At that point, I thought to myself that my credulity had limits and that no one could get me to believe that cigarette smoke could cure a sick child. On the contrary, I thought that blowing smoke on the child could only worsen its condition.

A few evenings later, during one of our taped conversations, I returned to this question:

"When one does a cure, like the one you did the other day for Sabino, how does the tobacco work? If you are the one who smokes it, how can it cure the person who does not smoke?"

"I always say, the property of tobacco is that it shows me the reality of things. I can see things as they are. And it gets rid of all the pains."

"Ah, but how did one discover this property? Does tobacco grow wild in the forest?"

"There is a place, for example in Napiari, where there are enormous quantities of tobacco growing."

"Where?"

"In the Perene. We found out about its power thanks to ayahuasca, that other plant, because it is the mother."

"Who is the mother, tobacco or ayahuasca?"

"Ayahuasca."

"And tobacco is its child?"

"It's the child."

"Because tobacco is less strong?"

"Less strong."

"You told me that ayahuasca and tobacco both contain God."

"That's it."

"And you said that souls like tobacco. Why?"

"Because tobacco has its method, its strength. It attracts the maninkari. It is the best contact for the life of a human being."

"And these souls, what are they like?"

"I know that any living soul, or any dead one, is like those radio waves flying around in the air."

"Where?"

"In the air. That means that you do not see them, but they are there, like radio waves. Once you turn on the radio, you can pick them up. It's like that with souls; with ayahuasca and tobacco, you can see them and hear them."

"And why is it that when one listens to the ayahuasquero singing, one hears music like one has never heard before, such beautiful music?"

"Well, it attracts the spirits, and as I have always said, if one thinks about it closely . . . [long silence]. It's like a tape recorder, you put it there, you turn it on, and already it starts singing: hum, hum, hum, hum, hum. You start singing along with it, and once you sing, you understand them. You can follow their music because you have heard their voice. So, it occurs, and one can see, like the last time when Ruperto was singing."

As I LISTENED to these explanations, I realized that I did not really believe in the existence of spirits. From my point of view, spirits were at best metaphors. Carlos, on the other hand, considered spirits to be firmly rooted in the material world, craving tobacco, flying like radio waves, and singing like tape recorders. So my attitude was ambiguous. On the one hand, I wanted to understand what Carlos thought, but on the other, I couldn't take what he said seriously because I did not believe it.

This ambiguity was reinforced by what people said about spirits; namely, that contact with spirits gave one power not only to cure, but to cause harm.

One evening I accompanied Carlos and Ruperto to the house

of a third man, whom I will call M. Word had gone around that Ruperto, just back from an eight-year absence, had learned his lessons well with the Shipibo ayahuasqueros. For his part, M. boasted that he had a certain experience with hallucinogens, and said that he was curious to see just how good Ruperto was.

M. lived on the crest of a little hill surrounded by forest. We arrived at his house around eight in the evening. After the customary greetings, we sat down on the ground. Ruperto produced his bottle of ayahuasca and placed it at the bottom of the ladder leading up to the house's platform, saying, "This is its place." Then he passed around a rolled cigarette and blew smoke on the bottle and on M. Meanwhile, Carlos took my hands and also blew smoke on them. The sweet smell of tobacco and the blowing feeling on my skin were pleasurable.

Three months had gone by since my first ayahuasca session. I felt physically relaxed, yet mentally apprehensive. Was I going to see terrifying snakes again? We drank the bitter liquid. It seemed to me that Ruperto filled my cup less than the others. I sat in silence. At one point, with my eyes closed, my body seemed to be very long. Ruperto started singing. M. accompanied him, but sang a different melody. The sound of this dissonant duo was compelling, though the rivalry between the two singers implied a certain tension. Carlos remained silent throughout.

I continued feeling calm. Apart from a few kaleidoscopic images, I did not have any particularly remarkable visions, nor did I feel nauseated. I started to think that I had not drunk enough ayahuasca. When Ruperto asked me whether I was "drunk," I answered "not yet." He asked me whether I would like some more. I told him that I was not sure and wanted perhaps to wait a bit. I asked Carlos in a whisper for his opinion. He advised me to wait.

I spent approximately three hours sitting on the ground in the dark in a slightly hypnotic, but certainly not hallucinatory state of mind. In the darkness, I could only make out the shape of the other participants. Both Carlos and M. had told Ruperto that they were "drunk."

The session came to a rather abrupt end. Carlos stood up and said with unusual haste that he was going home to rest. I got up to accompany him and thanked both our host and Ruperto, to whom I confided that I had been slightly fearful of the ayahuasca. He said, "I know, I saw it when we arrived."

Carlos and I had only one flashlight. He took it and guided us along the path through the forest. I followed him closely to take full advantage of the beam. After covering approximately three hundred yards, Carlos suddenly yelped and scratched at the back of his calf, from which he seemed to extract some kind of sting. In the confusion, what he was holding between his fingers fell to the ground. He said, "That man is shameless. Now he is shooting his arrows at me." I was relieved to hear his words, because I was afraid a snake had bitten him, but I had no idea what he was talking about. I asked questions, but he interrupted, saying, "Later. Now, let's go." We marched over to his house.

On arrival, Carlos was visibly upset. He finally explained that M. had shot one of his arrows at him, "because he wants to dominate, and show that he is stronger."

For my part, I was left with a doubt. How could one really aim a little sting in total darkness across three hundred yards of forest and touch the back of the calf of a person walking in front of someone else?

Nevertheless, Carlos was ill the following day, and the tension between him and M. continued to the end of my stay in Quirishari.

These suspicions of sorcery gave rise to a network of rumors and counterrumors that partially undermined the community's atmosphere of goodwill.

Contact with the spirits may allow one to learn about the medicinal properties of plants and to cure. But it also gives the possibility of exploiting a destructive energy. According to the practitioners of shamanism, the source of knowledge and power to which they gain access is double-edged.

TOWARD THE END of my stay in Quirishari, I read over my fieldnotes and drew up a long list of questions. Most of them concerned the central subject of my investigation, but several dealt with the shamanic and mythological elements that had mystified me. In one of my last taped conversations with Carlos, I asked him about these matters:

> "Are tabaquero and ayahuasquero the same?"
>
> "The same."
>
> "Good, and I also wanted to know why it is that one sees snakes when one drinks ayahuasca."
>
> "It's because the mother of ayahuasca is a snake. As you can see, they have the same shape."
>
> "But I thought that ayahuasca was the mother of tobacco?"
>
> "That's right."
>
> "So who is the true owner of these plants, then?"
>
> "The owner of these plants, in truth, is like God; it is the maninkari. They are the ones who help us. Their existence knows neither end nor illness. That's why they say when the ayahuasquero puts his head into the dark room: 'If you want me to help you, then you must do things well, I will give you the power not for your personal gain, but for the good of all.' So

clearly, that is where the force lies. It is by believing the plant that you will have more life. That is the path. That's why they say that there is a very narrow path on which no one can travel, not even with a machete. It is not a straight path, but it is a path nonetheless. I hold on to those words and to the ones that say that truth is not for sale, that wisdom is for you, but it is for sharing. Translating this, it means it is bad to make a business of it."

During my last interviews with Carlos, I had the impression that the more I asked questions, the less I understood his answers. Not only was ayahuasca the mother of tobacco, which I already knew, but the mother of ayahuasca was a snake. What could this possibly mean—other than that the mother of the mother of tobacco is a snake?

On leaving Quirishari, I knew I had not solved the enigma of the hallucinatory origin of Ashaninca ecological knowledge. I had done my best, however, to listen to what people said. I had constantly tried to reduce the nuisance of my presence as an anthropologist. I never took notes in front of people to avoid their feeling spied on. Mostly, I would write in the evening, lying on my blanket, before going to sleep. I would simply note what I had done during the day and the important things that people had said. I even tried thinking about my presuppositions, knowing that it was important to objectify my objectifying gaze. But the mystery remained intact.

I left with the strange feeling that the problem had more to do with my incapacity to understand what people had said, rather than the inadequacy of their explanations. They had always used such simple words.

ENIGMA IN RIO

I n late 1986, I went home to rural Switzerland to write my dissertation. Two years later, after becoming a "doctor of anthropology," I felt compelled to put my ideas to practice. Under Ashaninca influence I had come to consider that practice was the most advanced form of theory. I was tired of doing research. Now I wanted to act. So I turned my back on the enigma of plant communication.

I started working for Nouvelle Planète, a small Swiss organization that promotes community development in Third World countries. In 1989, I traveled around the Amazon Basin, talking with indigenous organizations and collecting projects for the legal recognition of indigenous territories. Then I gathered funds for these projects in Europe.

This took up my time for four years. Most of the projects that I presented to individuals, communities, citizen groups, foundations, and even a governmental organization were funded and carried out successfully.[1]

Working hand in hand with indigenous organizations, South American topographers and anthropologists did the actual job of land titling. Each country has a different set of laws specifying the requirements for official recognition of indigenous territories. In Peru, for example, topographers must visit and map in detail the rivers, forests, mountains, fields, and villages used by a given indigenous people, and anthropologists must account for the number of persons occupying the territory and describe their way of life; these documents are then registered with the Ministry of Agriculture, which processes them and issues official land titles in the names of the indigenous communities. These titles guarantee the collective territorial ownership by people who have occupied the land for millennia, in many cases.

The funds that I raised served to pay the salaries of the anthropologists and topographers, their travel expenses in isolated parts of the rainforest, the materials needed for mapmaking, and the cost of following the documents through the bureaucratic process. The most successful project was carried out in Peru's Putumayo, Napo, and Ampiyacu regions by AIDESEP, the national federation of indigenous organizations of the Peruvian Amazon; they hired the topographers and anthropologists and managed to gain titles to close to one and a half million acres of land for only U.S.$21,525.

Part of my work consisted of flying to South America occasionally, visiting the areas that had been titled, and checking the accounts. Given the difficulties indigenous people often have learning accounting, I was surprised to find in most cases that things had been done according to the plans laid out in the initial projects.

Back in Europe, I would give talks explaining why it makes ecological sense to demarcate the territories of indigenous people in the Amazonian rainforest, saying that they alone know how to use it sustainably. I would point out the *rational* nature of indigenous agricultural techniques such as polyculture and the use of small clearings. The more I talked, however, the more I realized that I was not telling the whole truth as I understood it.

I was *not* saying that these Amazonian people claim that their botanical knowledge comes from plant-induced hallucinations; I had tried these hallucinogens myself under their supervision, and my encounter with the fluorescent snakes had modified my way of looking at reality. In my hallucinations, I had learned important things—that I am just a human being, for example, and am intimately linked to other life forms and that true reality is more complex than our eyes lead us to believe.

I did not talk about these things, because I was afraid people would not take me seriously.

At this point, "being taken seriously" had to do with effective fund-raising more than with the fear of disqualification from an academic career.

In June 1992 I went to Rio to attend the world conference on development and environment. At the "Earth Summit," as it was known, everybody had suddenly become aware of the ecological knowledge of indigenous people. The governments of the world mentioned it in every treaty they signed[2]; personal care and pharmaceutical companies talked of marketing the natural products of indigenous people at "equitable" prices.[3] Meanwhile, ethnobotanists and anthropologists advanced impressive numbers relative to the intellectual property rights of indigenous people: 74 percent of the modern pharmacopoeia's plant-based remedies were first discovered by "traditional" societies; to this day, less than 2 percent of all plant species have been fully tested in laboratories, and the great majority of the remaining 98 percent are in tropical forests; the Amazon contains half of all the plant species on Earth[4]; and so on.

In Rio the industrial and political worlds were just waking up to the economic potential of tropical plants. The biotechnology of the 1980s had opened up new possibilities for the exploitation of natural resources. The biodiversity of tropical forests suddenly represented a fabulous source of unexploited wealth, but without the botanical knowledge of indigenous people, biotechnicians would be reduced to testing blindly the medicinal properties of the world's estimated 250,000 plant species.[5]

Indigenous people let their position on the matter be known during their own conference, held on the outskirts of Rio a week before the official summit. Following the lead of the Amazonian delegates, they declared their opposition to the Convention on Biodiversity that governments were about to sign, because it lacked a concrete mechanism to guarantee the compensation of their botanical knowledge. The Amazonian representatives based their position on experience: Pharmaceutical companies have a history of going to the Amazon to sample indigenous plant remedies and then of returning to their laboratories to synthesize and patent the active ingredients without leaving anything for those who made the original discovery.[6]

CURARE is the best-known example of this kind of poaching. Several millennia ago, Amazonian hunters developed this muscle-paralyzing substance as a blow-gun poison. It kills tree-borne animals without poisoning the meat while causing them to relax their grip and fall to the ground. Monkeys, when hit with an untreated arrow, tend to wrap their tails around branches and die out of the archer's reach. In the 1940s, scientists realized that curare could greatly facilitate surgery of the torso and of the vital organs, because it interrupts nerve impulses and relaxes all muscles, including breathing muscles. Chemists synthesized derivatives of the plant mixture by modifying the molecular structure of one of its active ingredients. Currently, anesthesiologists who "curarize" their patients use only synthetic compounds. In the entire process, everyone has received compensation for their work except the developers of the original product.[7]

Most of the time scientists balk at recognizing that "Stone Age Indians" could have developed anything. According to the usual theory, Indians stumbled on nature's useful molecules by chance experimentation. In the case of curare, this explanation seems improbable. There are forty types of curares in the Amazon, made from seventy plant species. The kind used in modern medicine comes from the Western Amazon. To produce it, it is necessary to combine several plants and boil them for seventy-two hours, while avoiding the fragrant but mortal vapors emitted by the broth. The final product is a paste that is inactive unless injected under the skin. If swallowed, it has no effect.[8] It is difficult to see how anybody could have stumbled on this recipe by chance experimentation.

Besides, how could hunters in the tropical forest, concerned with preserving the quality of the meat, have even imagined an intravenous solution? When one asks these people about the invention of curare, they almost invariably answer that it has a mythical origin. The Tukano of the Colombian Amazon say that the creator of the universe invented curare and gave it to them.[9]

IN RIO, ethnobotanists often cited the example of curare to demonstrate that the knowledge of Amazonian people had already contributed significantly to the development of medical science. They also discussed other plants of the indigenous pharmacopoeia that had only recently started to interest scientists: An extract of the *Pilocarpus jaborandi* bush used by the Kayapo and the Guajajara had recently been turned into a glaucoma remedy by Merck, the multinational pharmaceutical company, which was also devising a new anticoagulant based on the *tikiuba* plant of the Uru-eu-Wau-Wau. The fruit of *Couroupita guienensis* used by the Achuar to treat fungal infections, and the leaves of the *Aristolochia* vine brewed into a tea by the Tirio for the relief of stomachache, also attracted interest, along with many other unidentified plants that indigenous Amazonians use to cure skin lesions, diarrhea, snakebite, and so on. [10]

At the Earth Summit, everybody was talking about the ecological knowledge of indigenous people, but certainly no one was talking about the hallucinatory origin of some of it, as claimed by the indigenous people themselves. Admittedly, most anthropologists and ethnobotanists did not know about it, but even those who did said nothing, presumably because there is no way to do so and be taken seriously. Colleagues might ask, "You mean Indians claim they get molecularly verifiable information from their hallucinations? You don't take them literally, do you?" What could one answer?

It is true that not all of the world's indigenous people use hallucinogenic plants. Even in the Amazon, there are forms of shamanism based on techniques other than the ingestion of hallucinogens; but in Western Amazonia, which includes the Peruvian, Ecuadorian, and Colombian part of the basin, it is hard to find a culture that does not use an entire panoply of psychoactive plants. According to one inventory, there are seventy-two ayahuasca-using cultures in Western Amazonia.[11]

Richard Evans Schultes, the foremost ethnobotanist of the twentieth century, writes about the healers of a region in Colombia that he considers to be one of the centers of Western Amazonian shamanism: "The medicine men of the Kamsá and Inga tribes of the Valley of Sibundoy have an unusually extensive knowledge of medicinal and toxic plants. . . . One of the most renowned is Salvador Chindoy, who insists that his knowledge of the medicinal value of plants has been taught to him by the plants themselves through the hallucinations he has experienced in his long lifetime as a medicine man."[12]

Schultes does not say anything further about the hallucinatory origin of the botanical expertise of Amazonian people, because there is nothing one can say without contradicting two fundamental principles of Western knowledge.

First, hallucinations *cannot* be the source of real information, because to consider them as such is the definition of psychosis. Western knowledge considers hallucinations to be at best illusions, at worst morbid phenomena.[13]

Second, plants do not communicate like human beings. Scientific theories of communication consider that only human beings use abstract symbols like words and pictures and that plants do not relay information in the form of mental images.[14] For science, the human brain is the source of hallucinations, which psychoactive plants merely trigger by way of the hallucinogenic molecules they contain.

It was in Rio that I realized the extent of the dilemma posed by the hallucinatory knowledge of indigenous people. On the one hand, its results are empirically confirmed and used by the pharmaceutical industry; on the other hand, its origin cannot be discussed scientifically because it contradicts the axioms of Western knowledge.

When I understood that the enigma of plant communication was a blind spot for science, I felt the call to conduct an in-depth investigation of the subject. Furthermore, I had been carrying the mystery of plant communication around since my stay with the

Ashaninca, and I knew that explorations of contradictions in science often yield fruitful results. Finally, it seemed to me that the establishment of a serious dialogue with indigenous people on ecology and botany required that this question be addressed.

AFTER RIO, I knew that I wanted to write a book on the subject. My original intention was simply to name the enigma and to establish an exploratory map of the following cul-de-sac, or paradox: We can use their knowledge, but as soon as we reach the question of its origin, we must turn back.

By drinking ayahuasca in Quirishari, I had gone beyond the signs saying "you have reached the limits of science" and had found an irrational and subjective territory that was terrifying, yet filled with information. So I knew that the cul-de-sac had a passage that is normally hidden from the rational gaze and that leads to a world of surprising power.

However, I did not imagine for an instant that I could solve the enigma. I was convinced that I was dealing with an essentially paradoxical phenomenon that was not subject to solution.

DEFOCALIZING

Twelve months after the Rio conference a publisher accepted my proposal for a book on Amazonian shamanism and ecology. I was going to call it *Ecological hallucinations*. Several weeks later my employer agreed to let me spend part of my time working on the book.

I was set to investigate the enigma of plant communication. But where was I to begin?

My initial impulse would have been to return to the Peruvian Amazon and spend some time with the ayahuasqueros. However, my life had changed. I was no longer a free-roaming anthropologist, but the father of two young children. I was going to have to conduct my investigation from my office and the nearest library, rather than from the forests of Peru.

I started by rereading my fieldnotes and the transcripts of the Carlos Perez Shuma interviews. I paid particular attention to the strange passages I had left out of my thesis. Then, given that writing is an extension of thinking, I drafted a preliminary version of a first chapter on my arrival in Quirishari and my initial ayahuasca experience.

During this immersion in mysterious moments of my past, I started thinking about what Carlos had said. What if I took him literally? What if it were true that nature speaks in signs and that the secret to understanding its language consists in noticing similarities in shape or in form? I liked this idea and decided to read the anthropological texts on shamanism paying attention not only to their content but to their style. I taped a note on the wall of my office: "Look at the FORM."

One thing became clear as I thought back to my stay in Quirishari. Every time I had doubted one of my consultants' explanations, my understanding of the Ashaninca view of reality had seized up; conversely, on the rare occasions that I had managed to silence my doubts, my understanding of local reality had been enhanced—as if there were times when one had to believe in order to see, rather than the other way around.

This realization led me to decide, now that I was trying to map the cul-de-sac of hallucinatory knowledge, that it would be useful not only to establish its limits from a rational perspective, but to suspend disbelief and note with equal seriousness the outline of the ayahuasqueros' notions on the other side of the apparent impasse.

I read for weeks. I started by refreshing my memory and going over the basic texts of anthropology as well as the discipline's new, self-critical vein. Then I devoured the literature on shamanism, which was new to me. I had not read as much since my doctoral examinations nine years previously and was pleased to rediscover this purely abstract level of reality. With an enthusiasm that I never had at university I took hundreds of pages of reading notes, which I then categorized.

Five months into my investigation, my wife and I visited friends who introduced us during the evening to a book containing colorful "three-dimensional images" made up of seemingly disordered dots. To see a coherent and "3-D" image emerge from the blur, one had to defocalize one's gaze. "Let your eyes go," our hostess told me, "as if you were looking through the book without seeing it. Relax into the blur and be patient." After several attempts, and seemingly by magic, a remarkably deep stereogram sprang out of the page that I was holding in front of me. It showed a dolphin leaping in the waves. As soon as I focused normally on the page, the dolphin disappeared, along with the waves in front of it and behind it, and all I could see were muddled dots again.

This experience reminded me of Bourdieu's phrase "to objectify one's objectifying relationship," which is another way of saying "to become aware of one's gaze." That is precisely what one had to do in order to see the stereogram. This made me think that my dissatisfaction with the anthropological studies of shamanism was perhaps due to the necessarily focalized perspective of academic anthropologists, who failed to grasp shamanic phenomena in the same way that the normal gaze failed to see "three-dimensional images." Was there perhaps a way of relaxing one's gaze and seeing shamanism more clearly?

During the following weeks I continued reading, while trying to relax my gaze and pay attention to the texts' style, as much as to their content. Then I started writing a preliminary version of a second chapter on anthropology and shamanism. One afternoon, as I was writing, I suddenly saw a strikingly coherent image emerge from the muddle, as in a stereogram: Most anthropologists who had studied shamanism had only seen their own shadow. This went for the schizophrenics, the creators of order, the jacks-of-all-trades, and the creators of disorder.

This vision shook me. I felt that I had finally found a warm trail. Without wasting time, I continued in the same direction.

As I felt certain that the enigma of hallucinatory knowledge was only an apparent dead end, and as I was trying to suspend disbelief, I started wondering whether I might not be able to find a solution after all. The passage that led to the shamanic world was certainly hidden from normal vision, but perhaps there was a way of perceiving it stereoscopically. . . .

Speculating in this way, I realized that the hallucinations I had seen in Quirishari could also be described as three-dimensional images invisible to a normal gaze. According to my Ashaninca friends, it was precisely by reaching the hallucinatory state of consciousness that one crossed the impasse. For them, there was no fundamental contradiction between the practical reality of their life in the rainforest and the invisible and irrational world of ayahuasqueros. On the contrary, it was by going back and forth between these two levels that one could bring back useful and verifiable knowledge that was otherwise unobtainable. This proved to me that it was possible to reconcile these two apparently distinct worlds.

I also felt that I needed to improve my defocalization skills in order to succeed. I live not far from a castle that belonged to the family of Arthur Conan Doyle, the author of the Sherlock Holmes investigations. During my youth, I had often admired the famous detective's "lateral" methods, where he would lock himself into his office and play discordant tunes on his violin late into the night—to emerge with the key to the mystery. In the wintry fogs of the Swiss plateau, I started following Holmes's example. Once the children were in bed, I would go down to my office and get to work with hypnotically dissonant music playing in the background.[1]

Some evenings I would go further. Given that walking makes thinking easier, I would dress up warmly and go for strolls in the misty darkness with my tape recorder. Accompanied only by the rhythm of my boot heels, I would think aloud about all the imaginable solutions to the enigma that was beginning to obsess me. The following day I would transcribe these nebulous soliloquies looking for new perspectives. Some passages truly helped me understand where I was trying to go: "You must defocalize your gaze so as to perceive science and the indigenous vision at the same time. Then the common ground between the two will appear in the form of a stereogram. . . ."

My social life became nonexistent. Apart from a few hours in the afternoon with my children, I spent most of my time reading and thinking. My wife started saying I was absent even when I was in the room. She was right, and I could not hear her because I was obsessed. The more I advanced with this unusual methodology, the fresher the trail seemed.

FOR SEVERAL WEEKS, I went over the scientific literature on hallucinogens and their supposed effects on the human brain.

Here is a fact I learned during my reading: We do not know how our visual system works. As you read these words, you do not *really* see the ink, the paper, your hands, and the surroundings, but an internal and three-dimensional image that reproduces them almost exactly and that is constructed by your brain. The photons reflected by this page strike the retinas of your eyes, which transform them into electrochemical information; the optic nerves relay this information to the visual cortex at the back of the head, where a cascade-like network of nerve cells separates the input into categories (form, color, movement, depth, etc.). How the brain goes about reuniting these sets of categorized information into a coherent image is still a mystery. This also means that the neurological basis of consciousness is unknown.[2]

If we do not know how we see a real object in front of us, we understand even less how we perceive something that is not there. When a person hallucinates, there is no external source of visual stimulation, which, of course, is why cameras do not pick up hallucinatory images.

Strangely, and with few exceptions, these basic facts are not mentioned in the thousands of scientific studies on hallucinations; in books with titles such as *Origin and mechanisms of hallucinations,* experts provide partial and mainly hypothetical answers, which they formulate in complicated terms, giving the impression that they have attained the objective truth, or are about to do so.[3]

The neurological pathways of hallucinogens are better understood than the mechanisms of hallucinations. During the 1950s, researchers discovered that the chemical composition of most hallucinogens closely resembles that of serotonin, a hormone produced by the human brain and used as a chemical messenger between brain cells. They hypothesized that hallucinogens act on consciousness by fitting into the same cerebral receptors as serotonin, "like similar keys fitting the same lock."[4]

LSD, a synthetic compound unknown in nature, does not have the same profile as the organic molecules such as dimethyltrypta-

| Serotonin (brain hormone) | Psilocybin psilocin (organic hallucinogen) | N, N- Dimethyltryptamine (organic hallucinogen and brain hormone) | LSD (inorganic hallucinogen) |

mine or psilocybin. Nevertheless, the great majority of clinical investigations focused on LSD, which was considered to be the most powerful of all hallucinogens, given that only 50-millionths of a gram brings on its effects.[5]

In the second half of the 1960s, hallucinogens became illegal in the Western world. Shortly thereafter, scientific studies of these substances, which had been so prolific during the previous two decades, were stopped across the board. Ironically it was around this time that several researchers pointed out that, according to science's strict criteria, LSD most often does not induce true hallucinations, where the images are confused with reality. People under the influence of LSD nearly always know that the visual distortions or the cascades of dots and colors that they perceive are not real, but are due to the action of a psychedelic agent. In this sense, LSD is "pseudo-hallucinogenic."[6]

So the scientific studies of hallucinogens focused mainly on a product that is not really hallucinogenic; researchers neglected the natural substances, which have been used for thousands of years by hundreds of peoples, in favor of a synthetic compound invented in a twentieth-century laboratory.[7]

In 1979, it was discovered that the human brain seems to secrete dimethyltryptamine—which is also one of the active ingredients of ayahuasca. This substance produces true hallucinations, in which the visions replace normal reality convincingly, such as fluorescent snakes to whom one excuses oneself as one steps over them. Unfortunately, scientific research on dimethyltryptamine is rare. To this day, the clinical studies of its effects on "normal" human beings can be counted on the fingers of one hand.[8]

As I READ, the seasons turned. Suddenly winter gave way to spring, and the days began getting longer. I had just spent six full months focusing on other people's writings. Now I felt the time had come to pause momentarily, and then to start writing my book.

Making the most of the first warm spell of the year, I took a day off and went walking in a nature reserve with my tape recorder. The buds were starting to open, springs were gushing everywhere, and I was hoping that my ideas would do the same.

It had become clear to me that ayahuasqueros were somehow gaining access in their visions to verifiable information about plant properties. Therefore, I reasoned, the enigma of hallucinatory knowledge could be reduced to one question: Was this information coming from *inside* the human brain, as the scientific point of view would have it, or from the *outside* world of plants, as shamans claimed?

Both of these perspectives seemed to present advantages and drawbacks.

On the one hand, the similarity between the molecular profiles of the natural hallucinogens and of serotonin seemed well and truly to indicate that these substances work like keys fitting into the same lock *inside* the brain. However, I could not agree with the scientific position according to which hallucinations are merely discharges of images stocked in compartments of the subconscious memory. I was convinced that the enormous fluorescent snakes that I had seen thanks to ayahuasca did not correspond in any way to anything that I could have dreamed of, even in my most extreme nightmares. Furthermore, the speed and coherence of some of the hallucinatory images exceeded by many degrees the best rock videos, and I knew that I could not possibly have filmed them.[9]

On the other hand, I was finding it increasingly easy to suspend disbelief and consider the indigenous point of view as potentially correct. After all, there were all kinds of gaps and contradictions in the scientific knowledge of hallucinogens, which had at first seemed so reliable: Scientists do not know how these substances affect our consciousness, nor have they studied true hallucinogens in any detail. It no longer seemed unreasonable to me to consider that the information about the molecular content of plants could truly come from the plants themselves, just as ayahuasqueros claimed. However, I failed to see how this could work concretely.

With these thoughts in mind, I interrupted my stroll and sat down, resting my back against a big tree. Then I tried to enter into communication with it. I closed my eyes and breathed in the damp vegetal scent in the air. I waited for a form of communication to appear on my mental screen—but I ended up perceiving nothing more than the agreeable feeling of immersion in sunshine and fertile nature.

After about ten minutes, I stood up and resumed walking. Suddenly my thoughts turned again to stereograms. Maybe I would find the answer by looking at both perspectives simultaneously, with one eye on science and the other on shamanism. The solution would therefore consist in posing the question differently: It was not a matter of asking whether the source of hallucinations is internal *or* external, but of considering that it might be both at the same time. I could not see how this idea would work in practice, but I liked it because it reconciled two points of view that were apparently divergent.

The path I was following led to a crystalline cascade gushing out of a limestone cliff. The water was sparkling and tasted like champagne.

THE NEXT DAY I returned to my office with renewed energy. All I had to do was classify my reading notes on Amazonian shamanism and then I could start writing. However, before getting down to this task, I decided to spend a day following my fancy, freely paging through the piles of articles and notes that I had accumulated over the months.

In reading the literature on Amazonian shamanism, I had noticed that the personal experience of anthropologists with indigenous hallucinogens was a gray zone. I knew the problem well for having skirted around it myself in my own writings. One of the categories in my reading notes was called "Anthropologists and Ayahuasca." I consulted the card corresponding to this category, which I had filled out over the course of my investigation, and noted that the first subjective description of an ayahuasca experience by an anthropologist was published in 1968—whereas several botanists had written up similar experiences a hundred years previously.[10]

The anthropologist in question was Michael Harner. He had devoted ten lines to his own experience in the middle of an academic article: "For several hours after drinking the brew, I found myself, although awake, in a world literally beyond my wildest dreams. I met bird-headed people, as well as dragon-like creatures who explained that they were the true gods of this world. I enlisted the services of other spirit helpers in attempting to fly through the far reaches of the Galaxy. Transported into a trance where the supernatural seemed natural, I realized that anthropologists, including myself, had profoundly underestimated the importance of the drug in affecting native ideology."[11]

At first Michael Harner pursued an enviable career, teaching in reputable universities and editing a book on shamanism for Oxford University Press. Later, however, he alienated a good portion of his colleagues by publishing a popular manual on a series of shamanic techniques based on visualization and the use of drums.

One anthropologist called it "a project deserving criticism given M. Harner's total ignorance about shamanism."[12] In brief, Harner's work was generally discredited.

I must admit that I had assimilated some of these prejudices. At the beginning of my investigation, I had only read through Harner's manual quickly, simply noting that the first chapter contained a detailed description of his first ayahuasca experience, which took up ten pages this time, instead of ten lines. In fact, I had not paid particular attention to its content.

So, for pleasure and out of curiosity, I decided to go over Harner's account again. It was in reading this literally fantastic narrative that I stumbled on a key clue that was to change the course of my investigation.

Harner explains that in the early 1960s, he went to the Peruvian Amazon to study the culture of the Conibo Indians. After a year or so he had made little headway in understanding their religious system when the Conibo told him that if he really wanted to learn, he had to drink ayahuasca. Harner accepted not without fear, because the people had warned him that the experience was terrifying. The following evening, under the strict supervision of his indigenous friends, he drank the equivalent of a third of a bottle. After several minutes he found himself falling into a world of true hallucinations. After arriving in a celestial cavern where "a supernatural carnival of demons" was in full swing, he saw two strange boats floating through the air that combined to form "a huge dragon-headed prow, not unlike that of a Viking ship." On the deck, he could make out "large numbers of people with the heads of blue jays and the bodies of humans, not unlike the bird-headed gods of ancient Egyptian tomb paintings."

After multiple episodes, which would be too long to describe here, Harner became convinced that he was dying. He tried calling out to his Conibo friends for an antidote without managing to pronounce a word. Then he saw that his visions emanated from "giant reptilian creatures" resting at the lowest depths of his brain. These creatures began projecting scenes in front of his eyes, while informing him that this information was reserved for the dying and the dead: "First they showed me the planet Earth as it was eons ago, before there was any life on it. I saw an ocean, barren land, and a bright blue sky. Then black specks dropped from the sky by the hundreds and landed in front of me on the barren landscape. I could see the 'specks' were actually large, shiny, black creatures with stubby pterodactyl-like wings and huge whale-like bodies. . . . They explained to me in a kind of thought language that they were fleeing from something out in space. They had come to the planet Earth to escape their enemy. The creatures then showed me how they had created life on the planet in order to hide within the multitudinous forms and thus disguise their presence. Before me, the magnificence of plant and animal creation and speciation—hundreds of millions of years of activity—took place on a scale and with a vividness impossible to describe. I learned that the dragon-like creatures were thus inside all forms of life, including man."

At this point in his account, Harner writes in a footnote at the bottom of the page: "In retrospect one could say they were almost like DNA, although at that time, 1961, I knew nothing of DNA."[13]

I paused. I had not paid attention to this footnote previously. There was indeed DNA *inside* the human brain, as well as in the *outside* world of plants, given that the molecule of life containing genetic information is the same for all species. DNA could thus be considered a source of information that is both external and internal—in other words, precisely what I had been trying to imagine the previous day in the forest.

I plunged back into Harner's book, but found no further mention of DNA. However, a few pages on, Harner notes that "dragon" and "serpent" are synonymous. This made me think that the double helix of DNA resembled, in its *form,* two entwined serpents.

AFTER LUNCH, I returned to the office with a strange feeling. The reptilian creatures that Harner had seen in his brain reminded me of something, but I could not say what. It had to be a text that I had read and that was in one of the numerous piles of documents and notes spread out over the floor. I consulted the "Brain" pile, in which I had placed the articles on the neurological aspects of consciousness, but I found no trace of reptiles. After rummaging around for a while, I put my hand on an article called "Brain and mind in Desana shamanism" by Gerardo Reichel-Dolmatoff.

I had ordered a copy of this article from the library during my readings on the brain. Knowing from Reichel-Dolmatoff's numerous publications that the Desana of the Colombian Amazon were regular ayahuasca users, I had been curious to learn about their point of view on the physiology of consciousness. But the first time I had read the article, it had seemed rather esoteric, and I had relegated it to a secondary pile. This time, paging through it, I was stopped by a Desana drawing of a human brain with a snake lodged between the two hemispheres (*see top of page 57*).

I read the text above the drawing and learned that the Desana consider the fissure occupied by the reptile to be a "depression that was formed in the beginning of time (of mythical and embryological time) by the cosmic anaconda. Near the head of the serpent is a hexagonal rock crystal, just outside the brain; it is there where a particle of solar energy resides and irradiates the brain."[14]

Several pages further into the article, I came upon a second drawing, this time with two snakes (*see top of page 58*).

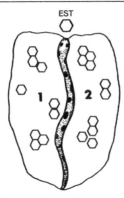

*The human brain. The left hemisphere is referred
to as Side One, and the right as Side Two. The
fissure is occupied by an anaconda.
(Redrawn from Desana sketches.)
From Reichel-Dolmatoff (1981, p. 81).*

According to Reichel-Dolmatoff, the drawing on page 58 shows that within the fissure "two intertwined snakes are lying, a giant anaconda (*Eunectes murinus*) and a rainbow boa (*Epicrates cenchria*), a large river snake of dark dull colors and an equally large land snake of spectacular bright colors. In Desana shamanism these two serpents symbolize a female and male principle, a mother and a father image, water and land . . . ; in brief, they represent a concept of binary opposition which has to be overcome in order to achieve individual awareness and integration. The snakes are imagined as spiralling rhythmically in a swaying motion from one side to another."[15]

Intrigued, I began reading Reichel-Dolmatoff's article from the beginning. In the first pages he provides a sketch of the Desana's main cosmological beliefs. My eyes stopped on the follow-

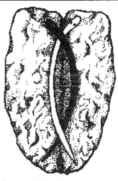

*The human brain. The fissure is occupied by an
anaconda and a rainbow boa.
(Redrawn from Desana sketches.)
From Reichel-Dolmatoff (1981, p. 88).*

ing sentence: "The Desana say that in the beginning of time their ancestors arrived in canoes shaped like huge serpents."[16]

At this point I began feeling astonished by the similarities between Harner's account, based on his hallucinogenic experience with the Conibo Indians in the Peruvian Amazon, and the shamanic and mythological concepts of an ayahuasca-using people living a thousand miles away in the Colombian Amazon. In both cases there were reptiles in the brain and serpent-shaped boats of cosmic origin that were vessels of life at the beginning of time. Pure coincidence?

To find out, I picked up a book about a third ayahuasca-using people, entitled (in French) *Vision, knowledge, power: Shamanism among the Yagua in the North-East of Peru.* This study by Jean-Pierre Chaumeil is, to my mind, one of the most rigorous on the subject. I started paging through it looking for passages relative to cosmological beliefs. First I found a "celestial serpent" in a drawing of the universe by a Yagua shaman. Then, a few pages away, another shaman is quoted as saying: "At the very beginning, before the birth of the earth, this earth here, our most distant ancestors lived on another earth. . . ." Chaumeil adds that the Yagua consider that all living beings were created by *twins,* who are "the two central characters in Yagua cosmogonic thought."[17]

These correspondences seemed very strange, and I did not know what to make of them. Or rather, I could see an easy way of interpreting them, but it contradicted my understanding of reality: A Western anthropologist like Harner drinks a strong dose of ayahuasca with one people and gains access, in the middle of the twentieth century, to a world that informs the "mythological" concepts of other peoples and allows them to communicate with life-creating spirits of cosmic origin possibly linked to DNA. This seemed highly improbable to me, if not impossible. However, I was getting used to suspending disbelief, and I had decided to follow my approach through to its logical conclusion. So I casually penciled in the margin of Chaumeil's text: "twins = DNA?"

These indirect and analogical connections between DNA and the hallucinatory and mythological spheres seemed amusing to me, or at most intriguing. Nevertheless, I started thinking that I had perhaps found with DNA the scientific concept on which to focus one eye, while focusing the other on the shamanism of Amazonian ayahuasqueros.

More concretely, I established a new category in my reading notes entitled "DNA Snakes."

Chapter 6

SEEING
CORRESPONDENCES

The following morning my wife and children left for a vacation in the mountains. I was going to be alone for ten days. I set about classifying my notes on the practices and beliefs of both indigenous and mestizo ayahuasqueros. This work took six days and revealed a number of constants across cultures.

Throughout Western Amazonia people drink ayahuasca at night, generally in complete darkness; beforehand, they abstain from sexual relations and fast, avoiding fats, alcohol, salt, sugar, and all other condiments. An experienced person usually leads the hallucinatory session, directing the visions with songs.[1]

In many regions, apprentice ayahuasqueros isolate themselves in the forest for long months and ingest huge quantities of hallucinogens. Their diet during this period consists mainly of bananas and fish, both of which are particularly rich in serotonin. It also happens that the long-term consumption of hallucinogens diminishes the concentration of this neurotransmitter in the brain. Most anthropologists are unaware of the biochemical aspect of this diet, however, and some go as far as to invent abstract explanations for what they call "irrational food taboos."[2]

As I classified my notes, I was looking out for new connections between shamanism and DNA. I had just received a letter from a friend who is a scientific journalist and who had read a preliminary version of my second chapter; he suggested that shamanism was perhaps "untranslatable into our logic for lack of corresponding concepts."[3] I understood what he meant, and I was trying to see precisely if DNA, without being exactly equivalent, might be the concept that would best translate what ayahuasqueros were talking about.

These shamans insist with disarming consistency on the existence of animate essences (or spirits, or mothers) which are common to all life forms. Among the Yaminahua of the Peruvian Amazon, for instance, Graham Townsley writes: "The central image dominating the whole field of Yaminahua shamanic knowledge is that of *yoshi*—spirit or animate essence. In Yaminahua thought all things in the world are animated and given their particular qualities by *yoshi*. Shamanic knowledge is, above all, knowledge of these entities, which are also the sources of all the powers that shamanism claims for itself. . . . it is through the idea of *yoshi* that the fundamental sameness of the human and the non-human takes shape."[4]

When I was in Quirishari, I already knew that the "animist" belief, according to which all living beings are animated by the same principle, had been confirmed by the discovery of DNA. I had learned in my high school biology classes that the molecule of life was the same for all species and that the genetic information in a rose, a bacterium, or a human being was coded in a universal language of four letters, A, G, C, and T, which are four chemical compounds contained in the DNA double helix.

So the rather obvious relationship between DNA and the animate essences perceived by ayahuasqueros was not new to me.

The classification of my reading notes did not reveal any further correspondences.

ON MY SEVENTH DAY of solitude, I decided to go to the nearest university library, because I wanted to follow up a last trail before getting down to writing: the trail of the life-creating twins that I had found in Yagua mythology.

As I browsed over the writings of authorities on mythology, I discovered with surprise that the theme of twin creator beings of celestial origin was extremely common in South America, and indeed throughout the world. The story that the Ashaninca tell about Avíreri and his sister, who created life by transformation, was just one among hundreds of variants on the theme of the "divine twins." Another example is the Aztecs' plumed serpent, Quetzalcoatl, who symbolizes the "sacred energy of life," and his twin brother Tezcatlipoca, both of whom are children of the cosmic serpent Coatlicue.[5]

I was sitting in the main reading room, surrounded by students, and browsing over Claude Lévi-Strauss's latest book, when I jumped. I had just read the following passage: "In Aztec, the word *coatl* means both 'serpent' and 'twin.' The name Quetzalcoatl can thus be interpreted either as 'Plumed serpent' or 'Magnificent twin.'"[6] A twin serpent, of cosmic origin, symbolizing the sacred energy of life? Among the Aztecs?

It was the middle of the afternoon. I needed to do some thinking. I left the library and started driving home. On the road back, I could not stop thinking about what I had just read. Staring out of the window, I wondered what all these twin beings in the creation myths of indigenous people could possibly mean.

When I arrived home, I went for a walk in the woods to clarify my thoughts. I started by recapitulating from the beginning: I was trying to keep one eye on DNA and the other on shamanism to discover the common ground between the two. I reviewed the correspondences that I had found so far. Then I walked in silence, because I was stuck. Ruminating over this mental block I recalled Carlos Perez Shuma's words: "Look at the FORM."

That morning, at the library, I had looked up DNA in several encyclopedias and had noted in passing that the shape of the double helix was most often described as a ladder, or a twisted rope ladder, or a spiral staircase. It was during the following split second, asking myself whether there were any ladders in shamanism, that the revelation occurred: "THE LADDERS! *The shamans' ladders, 'symbols of the profession' according to Métraux, present in shamanic themes around the world according to Eliade!*"

I rushed back to my office and plunged into Mircea Eliade's book *Shamanism: Archaic techniques of ecstasy* and discovered that there were "countless examples" of shamanic ladders on all five continents, here a "spiral ladder," there a "stairway" or "braided ropes." In Australia, Tibet, Nepal, Ancient Egypt, Africa, North and South America, "the symbolism of the rope, like that of the ladder, necessarily implies communication between sky and earth. It is by means of a rope or a ladder (as, too, by a vine, a bridge, a chain of arrows, etc.) that the gods descend to earth and men go up to the sky." Eliade even cites an example from the Old Testament, where Jacob dreams of a ladder reaching up to heaven, "with the angels of God ascending and descending on it." According to Eliade, the shamanic ladder is the earliest version of the idea of an *axis of the world*, which connects the different levels of the cosmos, and is found in numerous creation myths in the form of a tree.[7]

Until then, I had considered Eliade's work with suspicion, but suddenly I viewed it in a new light.[8] I started flipping through his other writings in my possession and discovered: *cosmic serpents*.

This time it was Australian Aborigines who considered that the creation of life was the work of a "cosmic personage related to universal fecundity, the Rainbow Snake," whose powers were symbolized by quartz crystals. It so happens that the Desana of the Colombian Amazon also associate the cosmic anaconda, creator of life, with a quartz crystal:

"The ancestral anaconda . . . guided by the divine rock crystal." From Reichel-Dolmatoff (1981, p. 79).

How could it be that Australian Aborigines, separated from the rest of humanity for 40,000 years, tell the same story about the creation of life by a cosmic serpent associated with a quartz crystal as is told by ayahuasca-drinking Amazonians?

The connections that I was beginning to perceive were blowing away the scope of my investigation. How could cosmic serpents from Australia possibly help my analysis of the uses of hallucinogens in Western Amazonia? Despite this doubt, I could not stop myself and charged ahead.

I seized the four volumes of Joseph Campbell's comparative work on world mythology. A German friend had given them to me at the beginning of my investigation, after I had told him about the book that I wanted to write. Initially, I had simply gone over the volume called *Primitive mythology*. I didn't much like the title, and the book neglected the Amazon Basin, not to mention hallucinogens. At the time, I had placed Campbell's masterpiece at the back of one of my bookshelves and had not consulted it further. I began paging through *Occidental mythology* looking for snakes. To my surprise I found one in the title of the first chapter. Turning the first page I came upon the following figure.

"The Serpent Lord Enthroned." From Campbell (1964, p. 11).

This figure is taken from a Mesopotamian seal of *c.* 2200 B.C. and shows "the deity in human form, enthroned, with his caduceus emblem behind and a fire altar before."[9] The symbol of this Serpent Lord was none other than a double helix. The similarity with the representation of DNA was unmistakable!

I feverishly paged through Campbell's books and found twisted serpents in most images representing sacred scenes. Campbell writes about this omnipresent snake symbolism: "Throughout the material in the Primitive, Oriental and Occidental volumes of this work, myths and rites of the serpent frequently appear, and in a remarkably consistent symbolic sense. Wherever nature is revered as self-moving, and so inherently divine, the serpent is revered as symbolic of its divine life."[10]

In Campbell's work I discovered a stunning number of creator gods represented in the form of a cosmic serpent, not only in Amazonia, Mexico, and Australia, but in Sumer, Egypt, Persia, India, the Pacific, Crete, Greece, and Scandinavia.

To check these facts, I consulted my French-language *Dictionary of symbols* under "serpent." I read: "It makes light of the sexes, and of the opposition of contraries; it is female and male too, *a twin to itself,* like so many of the important creator gods who are always, in their first representation, cosmic serpents. . . . Thus, the visible snake appears as merely the brief incarnation of a Great Invisible Serpent, which is causal and timeless, a master of the vital principle and of all the forces of nature. It is a primary *old god* found at the beginning of all cosmogonies, before monotheism and reason toppled it" (original italics).[11]

Campbell dwells on two crucial turning points for the cosmic serpent in world mythology. The first occurs "in the context of the patriarchy of the Iron Age Hebrews of the first millennium B.C., [where] the mythology adopted from the earlier neolithic and Bronze Age civilizations . . . became inverted, to render an argument just the opposite to that of its origin." In the Judeo-Christian creation story told in the first book of the Bible, one finds elements which are common to so many of the world's creation myths: the serpent, the tree, and the twin beings; but for the first time, the serpent, "who had been revered in the Levant for at least seven thousand years before the composition of the Book of Genesis," plays the part of the villain. Yahweh, who replaces it in the role of the creator, ends up defeating "the serpent of the cosmic sea, Leviathan."[12]

For Campbell, the second turning point occurs in Greek mythology, where Zeus was initially represented as a serpent; but around 500 B.C., the myths are changed, and Zeus becomes a serpent-killer. He secures the reign of the patriarchal gods of Mount Olympus by defeating Typhon, the enormous serpent-monster who is the child of the earth goddess Gaia and the incarnation of the forces of nature. Typhon "was so large that his head often knocked against the stars and his arms could extend from sunrise to sunset." In order to defeat Typhon, Zeus can count only on the help of Athene, "Reason," because all the other Olympians have fled in terror to Egypt.[13]

"Zeus against Typhon." From Campbell (1964, p. 239).

At this point, I wrote in my notes: "These patriarchal and exclusively masculine gods are incomplete as far as nature is concerned. DNA, like the cosmic serpent, is neither masculine nor feminine, even though its creatures are either one or the other, or both. Gaia, the Greek earth goddess, is as incomplete as Zeus. Like him, she is the result of the rational gaze, which separates before thinking, and is incapable of grasping the androgynous and double nature of the vital principle."

IT WAS PAST 8 P.M. and I had not eaten. My head was spinning in the face of the enormity of what I thought I was discovering. I decided to pause. I took a beer out of the refrigerator and put on some violin music. Then I started pacing up and down my office. What could all this possibly mean?

I turned on the tape recorder and tried answering my own question: "One, Western culture has cut itself off from the serpent/life principle, in other words DNA, since it adopted an exclusively rational point of view. Two, the peoples who practice what we call 'shamanism' communicate with DNA. Three, paradoxically, the part of humanity that cut itself off from the serpent managed to discover its material existence in a laboratory some three thousand years later.

"People use different techniques in different places to gain access to knowledge of the vital principle. In their visions shamans manage to take their consciousness down to the molecular level. This is precisely what Reichel-Dolmatoff describes, in his running commentary into the tape recorder of his own ayahuasca-induced visions ('like microphotographs of plants; like those microscopic stained sections; sometimes like from a pathology textbook'[14]).

"This is how they learn to combine brain hormones with monoamine oxidase inhibitors, or how they discover forty different sources of muscle paralyzers whereas science has only been able to imitate their molecules. When they say the recipe for curare was given to them by the beings who created life, they are talking literally. When they say their knowledge comes from beings they see in their hallucinations, their words mean exactly what they say.

"According to the shamans of the entire world, one establishes communication with spirits via music. For the ayahuasqueros, it is almost inconceivable to enter the world of spirits and remain silent. Angelika Gebhart-Sayer discusses the 'visual music' projected by the spirits in front of the shaman's eyes: It is made up of three-dimensional images that coalesce into sound and that the shaman imitates by emitting corresponding melodies.[15] I should check whether DNA emits sound or not.

"Another way of testing this idea would be to drink ayahuasca and observe the microscopic images. . . ."

As I said this, it dawned on me that I could approach this experience by looking at the book of paintings by Pablo Amaringo, a retired Peruvian ayahuasquero with a photographic memory.

These paintings are published in a book called *Ayahuasca visions: The religious iconography of a Peruvian shaman,* by Luis Eduardo Luna and Pablo Amaringo. In this book, Luna the anthropologist provides a mine of information on Amazonian shamanism, giving the context for fifty paintings of great beauty by Amaringo. The first time I saw these paintings, I was struck by their resemblance to my own ayahuasca-induced visions. According to Amaringo: "I only paint what I have seen, what I have experienced. I don't copy or take ideas for my paintings from any book." Luna says: "I have shown Pablo's paintings to several *vegetalistas,* and they have reacted with immediate interest and wonder—some have commented on how similar their own visions were to those depicted by Pablo, and some even recognize elements in them."[16]

On opening the book, I was stupefied to find all kinds of zigzag staircases, entwined vines, twisted snakes, and, above all, usually hidden in the margins, *double helixes,* like this one:

Several weeks later, I showed these paintings to a friend with a good understanding of molecular biology. He reacted in the same way as the *vegetalistas* to whom Luna had shown them: "Look, there's collagen. . . . And there, the axon's embryonic network with its neurites. . . . Those are triple helixes. . . . And that's DNA from afar, looking like a telephone cord. . . . This looks like chromosomes at a specific phase. . . . There's the spread-out form of DNA, and right next to it are DNA spools in their nucleosome structure."[17]

Even without these explanations, I was in shock. I quickly went over the index of *Ayahuasca visions*, but found no mention of either DNA, chromosomes, or double helixes.

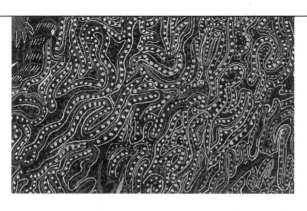

". . . the spread-out form of DNA . . ."

". . . chromosomes at a specific phase . . ."

". . . triple helixes of collagen . . ."

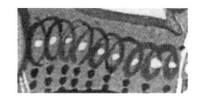

". . . DNA from afar, looking like a telephone cord . . ."

Was it possible that no one had noticed the molecular aspect of the images? Well, yes, because I myself had often admired them and showed them to people to explain what the hallucinatory sphere looked like, and I had not noticed them either. My gaze had been as focalized as usual. I had not been able to see simultaneously molecular biology and shamanism, which our rational mind separates—but which could well overlap and correspond.

I was staggered. It seemed that no one had noticed the possible links between the "myths" of "primitive peoples" and molecular biology. No one had seen that the double helix had symbolized the life principle for thousands of years around the world. On the contrary, everything was upside down. It was said that hallucinations could in no way constitute a source of knowledge, that Indians had found their useful molecules by chance experimentation, and that their "myths" were precisely myths, bearing no relationship to the real knowledge discovered in laboratories.

At this point I remembered Michael Harner's story. Had he not said that this information was reserved for the dead and the dying? Suddenly, I was overcome with fear and felt the urge to share these ideas with someone else. I picked up the phone and called an old friend, who is also a writer. I quickly took him through the correspondences I had found during the day: the twins, the cosmic serpents, Eliade's ladders, Campbell's double helixes, as well as Amaringo's. Then I added: "There is a last correlation that is slightly less clear than the others. The spirits one sees in hallucinations are three-dimensional, sound-emitting images, and they speak a language made of three-dimensional, sound-emitting images. In other words, they are made of their own language, like DNA."

There was a long silence on the other end of the line.

Then my friend said, "Yes, and like DNA they replicate themselves to relay their information." "Wait," I said, "I'm going to jot that down." "Precisely," he said, "instead of talking to me, you should be writing this down."[18]

I followed his advice, and it was in writing my notes on the relationship between the hallucinatory spirits made of language and DNA that I remembered the first verse of the first chapter of the Gospel according to John: "In the beginning was the *logos*"—the word, the verb, the language.

That night I had a hard time falling asleep.

THE NEXT DAY I had to attend a professional meeting that bore no relationship to my research. I took advantage of the train ride to put things into perspective. I was feeling strange. On the one hand, entire blocks of intuition were pushing me to believe that the connection between DNA and shamanism was real. On the other hand, I was aware that this vision contradicted certain academic ideas, and that the links I had found so far were insufficient to trouble a strictly rational point of view.

Nonetheless, gazing out the train window at a random sample of the the Western world, I could not avoid noticing a kind of separation between human beings and all other species. We cut ourselves off by living in cement blocks, moving around in glass-and-metal bubbles, and spending a good part of our time watching other human beings on television. Outside, the pale light of an April sun was shining down on a suburb. I opened a newspaper and all I could find were pictures of human beings and articles about their activities. There was not a single article about another species.

Sitting on the train, I measured the paradox confronting me. I was a resolutely Western individual, and yet I was starting to believe in ideas that were not receivable from a rational point of view. This meant that I was going to have to find out more about DNA. Up to this point, I had only found biological correspondences in shamanism. It remained to be seen whether the contrary was also true, and whether there were shamanic correspondences in biology. More precisely, I needed to see whether shamanic notions about spirits corresponded to scientific notions about DNA. Basically I had only covered, at best, half the distance.

Even though my bookshelves were well stocked in anthropology and ecology, I owned no books about DNA or molecular biology; but I knew a colleague trained in both chemistry and literature who was going to be able to help me on that count.

AFTER THE MEETING, toward the end of the afternoon, I went over to my colleague's house. He had generously allowed me to look through his books in his absence.

I entered his office, a big room with an entire wall occupied by bookshelves, turned on the light, and started browsing. The biology section contained, among others, *The double helix* by James Watson, the co-discoverer with Francis Crick of the structure of DNA. I flipped through this book, looking at the pictures with interest, and put it aside.

A little further along on the same shelf, I came upon a book by Francis Crick entitled *Life itself: Its origin and nature*. I pulled it out and looked at its cover—and could not believe my eyes. It showed an image of the earth, seen from space, with a rather indistinct object coming from the cosmos and landing on it.

Francis Crick, the Nobel Prize–winning co-discoverer of the structure of DNA, was suggesting that the molecule of life was of extraterrestrial origin—in the same way that the "animist"

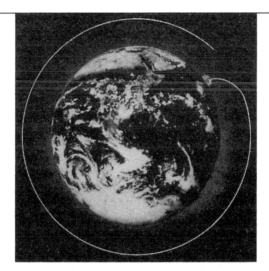

*Cover of Crick (1981), reproduced with permission
from Little, Brown & Co.*

peoples claimed that the vital principle was a serpent from the
cosmos.

I had never heard of Crick's hypothesis, called "directed
panspermia," but I knew that I had just found a new correspondence between science and the complex formed by shamanism
and mythology.

I sat down in the armchair and plunged into *Life itself: Its
origin and nature*.

CRICK, writing in the early 1980s, criticizes the usual scientific
theory on the origin of life, according to which a cell first appeared in the primitive soup through the random collisions of dis-
organized molecules. For Crick, this theory presents a major
drawback: It is based on ideas conceived in the nineteenth century, long before molecular biology revealed that the basic mech-

anisms of life are identical for all species and are extremely complex—and when one calculates the probability of chance producing such complexity, one ends up with inconceivably small numbers.

The DNA molecule, which excels at stocking and duplicating information, is incapable of building itself on its own. Proteins do this, but they are incapable of reproducing themselves without the information contained in the DNA. Life, therefore, is a seemingly inescapable synthesis of these two molecular systems. Moving beyond the famous question of the chicken and the egg, Crick calculates the probability of the chance emergence of one single protein (which could then go on to build the first DNA molecule). In all living species, proteins are made up of exactly the same 20 amino acids, which are small molecules. The average protein is a long chain made up of approximately 200 amino acids, chosen from those 20, and strung together in the right order. According to the laws of combinatorials, there is 1 chance in 20 multiplied by itself 200 times for a single specific protein to emerge fortuitously. This figure, which can be written 20^{200}, and which is roughly equivalent to 10^{260}, *is enormously greater than the number of atoms in the observable universe* (estimated at 10^{80}).

These numbers are inconceivable for a human mind. It is not possible to imagine all the atoms of the observable universe and even less a figure that is billions of billions of billions of billions of billions (etc.) times greater. However, since the beginning of life on earth, the number of amino acid chains that could have been synthesized by chance can only represent a minute fraction of all the possibilities. According to Crick: "The great majority of sequences can never have been synthesized at all, at any time. These calculations take account only of the amino acid sequence. They do not allow for the fact that many sequences would probably not fold up satisfactorily into a stable, compact shape. What fraction of all possible sequences would do this is not known, though it is surmised to be fairly small." Crick concludes that the organized complexity found at the cellular level "cannot have arisen by pure chance."

The earth has existed for approximately 4.5 billion years. In the beginning it was merely a radioactive aggregate with a surface temperature reaching the melting point of metal. Not really a hospitable place for life. Yet there are fossils of single-celled beings that are approximately 3.5 billion years old. The existence of a single cell necessarily implies the presence of DNA, with its 4-letter "alphabet" (A, G, C, T), and of proteins, with their 20-letter "alphabet" (the 20 amino acids), as well as a "translation mechanism" between the two—given that the instructions for the construction of proteins are coded in the language of DNA. Crick writes: "It is quite remarkable that such a mechanism exists at all and even more remarkable that every living cell, whether animal, plant or microbial, contains a version of it."[19]

Crick compares a protein to a paragraph made up of 200 letters lined up in the correct order. If the chances are infinitesimal for one *paragraph* to emerge in a billion years from a terrestrial soup, the probability of the fortuitous appearance, during the same period, of two *alphabets* and one *translation mechanism* is even smaller.

WHEN I LOOKED UP from Crick's book, it was dark outside. I was feeling both astonished and elated. Like a myopic detective bent over a magnifying glass while following a trail, I had fallen into a bottomless hole. For months I had been trying to untangle the enigma of the hallucinatory knowledge of Western Amazonia's indigenous people, stubbornly searching for the hidden passage in the apparent dead end. I had only detected the DNA trail two weeks previously in Harner's book. Since then I had mainly developed the hypothesis along intuitive lines. My goal was certainly not to build a new theory on the origin of life; but there I was—a poor anthropologist knowing barely how to swim, floating in a cosmic ocean filled with microscopic and bilingual serpents. I could see now that there might be links between science and shamanic, spiritual and mythological traditions, that seemed to have gone unnoticed, doubtless because of the fragmentation of Western knowledge.

With his book, Francis Crick provided a good example of this fragmentation. His mathematics were impeccable, and his reasoning crystalline; Crick was surely among twentieth-century rationality's finest. But he had not noticed that he was not the first to propose the idea of a snake-shaped vital principle of cosmic origin. All the peoples in the world who talk of a cosmic serpent have been saying as much for millennia. He had not seen it because the rational gaze is forever focalized and can examine only one thing at a time. It separates things to understand them, including the truly complementary. It is the gaze of the specialist, who sees the fine grain of a necessarily restricted field of vision. When Crick set about considering cosmogony from the serious perspective of molecular biology, he had long since put out of his analytical mind the myths of archaic peoples.

From my new point of view, Crick's scenario of "directed panspermia," in which a spaceship transports DNA in the form of frozen bacteria across the immensities of the cosmos, seemed less likely than the idea of an omniscient and terrifying cosmic serpent of unimaginable power. After all, life as described by Crick was based on a miniature language that had not changed a letter in four billion years, while multiplying itself in an extreme diversity

"A painting on hardboard of the Snake of the Marinbata people of Arnhem Land." From Huxley (1974, p. 127).

of species. The petals of a rose, Francis Crick's brain, and the coat of a virus are all built out of proteins made up of exactly the same 20 amino acids. A phenomenon capable of such creativity was surely not going to travel in a spaceship resembling those propelled containers imagined by human beings in the twentieth century.

This meant that the gaze of the Western specialist was too narrow to see the two pieces that fit together to resolve the puzzle. The distance between molecular biology and shamanism/mythology was an optical illusion produced by the rational gaze that separates things ahead of time, and as objectivism fails to objectify its objectifying relationship, it also finds it difficult to consider its presuppositions.

The puzzle to solve was: Who are we and where do we come from?

Lost in these thoughts, I started wondering about the cosmic serpent and its representation throughout the world. I walked over to the philosophy and religion sections in my colleague's library. Fairly rapidly I came across a book by Francis Huxley entitled *The way of the sacred*, filled with pictures of sacred images from around the world. I found a good number of images containing serpents or dragons, and in particular two representations of the Rainbow Snake drawn by Australian Aborigines. The first showed a pair of snakes zigzagging in the margins (*see top of page 78*).

The second was a rock painting of the Rainbow Snake. I looked at it more closely and saw two things: All around the serpent there were sorts of *chromosomes*, in their upside-down "U" shape, and underneath it there was a kind of *ladder*!

"A rock painting of the Walbiri tribe of Aborigines representing the Rainbow Snake." Photo by David Attenborough, from Huxley (1974, p. 126).

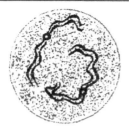 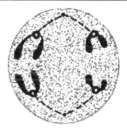

Early prophase: "Each chromosome is visible as two sister chromatids."   *Anaphase II: ". . . the migration of homologous chromatids to opposite poles."*

I rubbed my eyes, telling myself that I had to be imagining connections, but I could not get the ladder or the chromosomes to look like anything else.

Several weeks later I learned that U-shaped chromosomes were in "anaphase," one of the stages of cellular duplication, which is the central mechanism of the reproduction of life; and the first image of the zigzag snakes looks strikingly like chromosomes in the "early prophase," at the beginning of the same process.

However, I did not need this detail to feel certain now that the peoples who practice shamanism know about the hidden unity of nature, which molecular biology has confirmed, precisely because they have access to the reality of molecular biology.

It was at this point, in front of the picture of chromosomes painted by Australian Aborigines, that I sank into a fever of mind and soul that was to last for weeks, during which I floundered in dissonant mixes of myths and molecules.

Chapter 7

MYTHS
AND MOLECULES

First, I followed the mythological trail of the cosmic serpent, paying particular attention to its form. I found that it was often double:

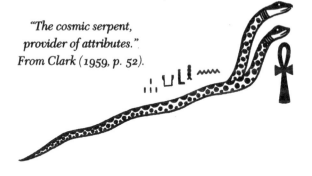

"The cosmic serpent, provider of attributes."
From Clark (1959, p. 52).

This Ancient Egyptian drawing does not represent a real animal, but a visual charade meaning "double serpent."

Quetzalcoatl, the Aztecs' plumed serpent, is not a real animal either. In living nature, snakes do not have arms or legs, and even less wings or feathers. A flying serpent is a contradiction in terms, a paradox, like a speaking mute. This is confirmed by the double etymology of the word *-coatl*, which means both "serpent" and "twin."

The Ancient Egyptians also represented the cosmic serpent with human feet.

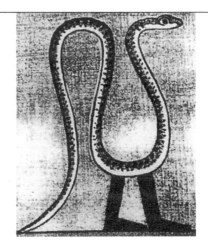

"Sito, the primordial serpent" (1300 B.C.)
From Clark (1959, p. 192).

Here, too, the image suggests that the primordial divinity is double, both serpent and "non-serpent."

In the early 1980s, ayahuasquero Luis Tangoa, living in a Shipibo-Conibo village in the Peruvian Amazon, offered to explain certain esoteric notions to anthropologist Angelika Gebhart-Sayer. Insisting that it was more appropriate to discuss these matters with images,[1] he made several sketches of the cosmic anaconda Ronín, including this one:

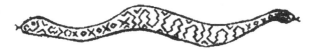

"Ronín, the two-headed serpent." From Gebhart-Sayer (1987, p. 42).

It would be possible to give many examples of double serpents of cosmic origin associated with the creation of life on earth, but it is important to avoid too strict an interpretation of these images, which can have several meanings at once. For instance, the wings of the serpent can signify both a paradoxical nature and a real ability to fly, in this case in the cosmos.

"The serpent of the earth becomes celestial; with wings, it can fly, and allows the mummy to ascend to the stars."
From Jacq (1993, p. 99).

Sometimes the winged serpent takes the form of a dragon, the mythical and double animal par excellence, which lives in the water and spits fire. According to the *Dictionary of symbols,* the dragon represents "the union of two opposed principles." Its androgynous nature is symbolized most clearly by the Ouroboros, the serpent-dragon, which "incarnates sexual union in itself, permanently self-fertilizing, as shown by the tail stuck in its mouth" (*see page 84*).

In living nature snakes do not bite their own tails. Nevertheless, the Ouroboros appears in some of the most ancient representations of the world, such as the bronze disk from Benin shown below. The *Dictionary of symbols* describes it as "doubtless the oldest African *imago mundi*, where its sinuous figure, associating opposites, encircles the primordial oceans in the middle of which floats the square of the earth below."[2]

"Here is the dragon that devours its tail."
From Maier (1965, p. 139).

"Ouroboros: bronze disk, Benin art." From Chevalier and
Gheerbrant (1982, p. 716).

Mythical serpents are often enormous. In the image from Benin, the Ouroboros surrounds the entire earth; in Greek mythology, the monster-serpent Typhon touches the stars with its head; and the first paragraph of the first chapter of Chuang-Tzu, the presumed founder of philosophical Taoism, describes an extremely long fish, inhabiting the celestial lake, that transforms itself into a bird and mounts spiraling into the sky. Chuang-Tzu says that the length of this cosmic fish-bird is "who knows how many thousand miles."[3]

Hindu mythology also provides an example of a serpent of immeasurable proportions, known as Sesha, the thousand-headed

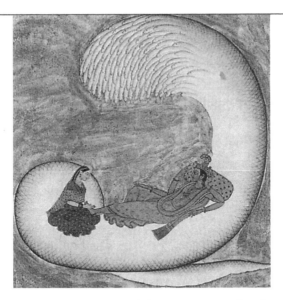

"Vishnu and his wife Lakshmi resting on Sesha, the thousand-headed serpent of eternity, in an interval between the cycles of creation." From Huxley (1974, pp. 188–189).

serpent that floats on the cosmic ocean while the twin creator beings Vishnu and Lakshmi recline in its coils.

Mythological serpents are almost invariably associated with water.[4] In the following drawing based on descriptions by ayahuasquero Laureano Ancon, the anaconda Ronín surrounds the entire earth, conceived as a "disc that swims in great waters"; Ronín itself is "half-submerged"—the anaconda being an aquatic species (*see top figure page 87*).

The cosmic serpent varies in size and nature. It can be small or large, single or double, and sometimes both at the same time (*see bottom figure page 87*). This picture was drawn by Luis Tangoa, who lives in the same village as Laureano Ancon. These two shamans would have had all the time in the world to reach an agreement about the appearance of the cosmic anaconda. Yet the former draws it as a single sperm *and* a two-headed snake, while the latter describes it as an anaconda of "normal" appearance that completely encircles the earth.

As the creator of life, the cosmic serpent is a master of metamorphosis. In the myths of the world where it plays a central part, it creates by transforming itself; it changes while remaining the same. So it is understandable that it should be represented differently at the same time.

I WENT ON TO LOOK for the connection between the cosmic serpent—the master of transformation of serpentine form that lives in water and can be both extremely long and small, single and double—and DNA. I found that DNA corresponds exactly to this description.

If one stretches out the DNA contained in the nucleus of a human cell, one obtains a two-yard-long thread that is only ten atoms wide. This thread is a billion times longer than its own

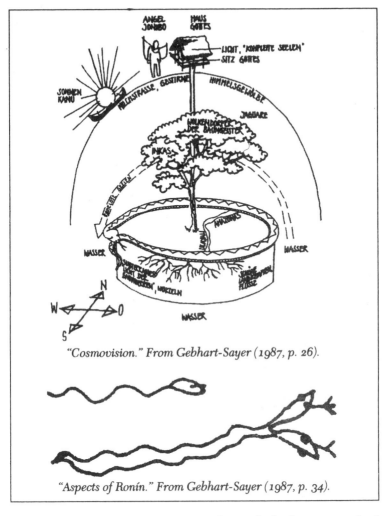

"Cosmovision." From Gebhart-Sayer (1987, p. 26).

"Aspects of Ronín." From Gebhart-Sayer (1987, p. 34).

width. Relatively speaking, it is as if your little finger stretched from Paris to Los Angeles.

A thread of DNA is much smaller than the visible light humans perceive. Even the most powerful optical microscopes cannot reveal it, because DNA is approximately 120 times narrower than the smallest wavelength of visible light.[5]

The nucleus of a cell is equivalent in volume to 2-millionths of a pinhead. The two-yard thread of DNA packs into this minute volume by coiling up endlessly on itself, thereby reconciling *extreme length* and *infinitesimal smallness,* like mythical serpents.

The average human being is made up of 100 thousand billion cells, according to some estimates. This means that there are approximately 125 billion miles of DNA in a human body—corresponding to 70 round-trips between Saturn and the Sun. You could travel your entire life in a Boeing 747 flying at top speed and you would not even cover one hundredth of this distance. Your personal DNA is long enough to wrap around the earth 5 million times.[6]

All the cells in the world contain DNA—be they animal, vegetal, or bacterial—and they are all filled with salt *water,* in which the concentration of salt is similar to that of the worldwide ocean. We cry and sweat what is essentially seawater. DNA bathes in water, which in turn plays a crucial role in establishing the double helix's shape. As DNA's four bases (adenine, guanine, cytosine, and thymine) are insoluble in water, they tuck themselves into the center of the molecule where they associate in pairs to form the rungs of the ladder; then they twist up into a spiraled stack to avoid contact with the surrounding water molecules. DNA's twisted ladder shape is a direct consequence of the cell's watery environment.[7] DNA goes together with water, just like mythical serpents do.

The DNA molecule is a *single* long chain made up of *two* interwoven ribbons that are connected by the four bases. These bases can only match up in specific pairs—A with T, G with C. Any other pairing of the bases is impossible, because of the arrangement of their individual atoms: A can bond only with T, G only with C. This means that one of the two ribbons is the back-to-front duplicate of the other and that the genetic text is *double:* It contains a main text on one of the ribbons, which is read in a precise direction by the transcription enzymes, and a backup text, which is inverted and most often not read.

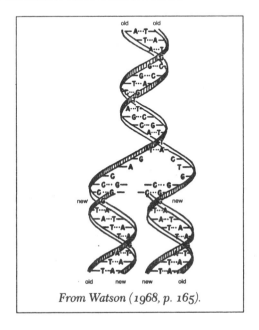

From Watson (1968, p. 165).

The second ribbon plays two essential roles. It allows the repair enzymes to reconstruct the main text in case of damage and, above all, it provides the mechanism for the duplication of the genetic message. It suffices to open the double helix as one might unzip a zipper, in order to obtain two separate and complementary ribbons that can then be rebuilt into double ribbons by the duplication enzymes. As the latter can place only an A opposite a T, and vice versa, and a G opposite a C, and vice versa, this leads to the formation of two *twin* double helixes, which are identical in every respect to the original. Twins are therefore central to life, just as ancient myths indicate, and they are associated with a serpentine form.

Without this copying mechanism, a cell would never be able to duplicate itself, and life would not exist.

DNA is the informational molecule of life, and its very essence consists in being *both single and double,* like the mythical serpents.

DNA AND ITS DUPLICATION MECHANISMS are the same for all living creatures. The only thing that changes from one species to another is the order of the letters. This constancy goes back to the very origins of life on earth. According to biologist Robert Pollack: "The planet's surface has changed many times over, but DNA and the cellular machinery for its replication have remained constant. Schrödinger's 'aperiodic crystal' understated DNA's stability: no stone, no mountain, no ocean, not even the sky above us, have been stable and constant for this long; nothing inanimate, no matter how complicated, has survived unchanged for a fraction of the time that DNA and its machinery of replication have coexisted."[8]

At the beginning of its existence, some 4.5 billion years ago, planet earth was an inhospitable place for life. As a molten lava fireball, its surface was radioactive; its water was so hot it existed only in the form of incondensable vapor, and its atmosphere, devoid of any breathable oxygen, contained poisonous gases such as cyanide and formaldehyde.

Approximately 3.9 billion years ago, the earth's surface cooled sufficiently to form a thin crust on top of the molten magma. Strangely, life, and thus DNA, appeared relatively quickly thereafter. Scientists have found traces of biological activity in sedimentary rocks that are 3.85 billion years old, and fossil hunters have found actual bacterial fossils that are 3.5 billion years old.

During the first 2 billion years of life on earth, the planet was inhabited only by anaerobic bacteria, for which oxygen is a poison. These bacteria lived in water, and some of them learned to use the hydrogen contained in the H_2O molecule while expelling the oxygen. This opened up new and more efficient metabolic pathways. The gradual enrichment of the atmosphere with oxygen allowed the appearance of a new kind of cell, capable of using oxygen and equipped with a nucleus for packing together its DNA. These nucleated cells are at least thirty times more voluminous than bacterial cells. According to biologists Lynn Margulis and Dorion Sagan: "The biological transition between bacteria and nucleated cells . . . is so sudden it cannot effectively be explained by gradual changes over time."

From that moment onward, life as we know it took shape. Nucleated cells joined together to form the first multicellular beings, such as algae. The latter also produce oxygen by photosynthesis. Atmospheric oxygen increased to about 21 percent and then stabilized at this level approximately 500 million years ago—thankfully, because if oxygen were a few percent higher, living beings would combust spontaneously. According to Margulis and Sagan, this state of affairs "gives the impression of a conscious decision to maintain balance between danger and opportunity, between risk and benefit."[9]

Around 550 million years ago, life exploded into a grand variety of multicellular species, algae and more complex plants and animals, living not only in water, but on land and in the air. Of all the species living at that time, not one has survived to this day. According to certain estimates, almost all of the species that have ever lived on earth have already disappeared, and there are between 3 million and 50 million species living currently.[10]

DNA is a *master of transformation,* just like mythical serpents. The cell-based life DNA informs made the air we breathe, the landscape we see, and the mind-boggling diversity of living beings of which we are a part. In 4 billion years, it has multiplied itself into an incalculable number of species, while remaining exactly the same.

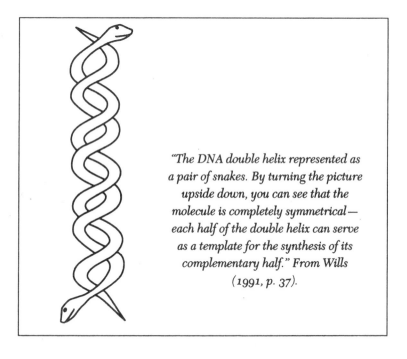

"The DNA double helix represented as a pair of snakes. By turning the picture upside down, you can see that the molecule is completely symmetrical — each half of the double helix can serve as a template for the synthesis of its complementary half." From Wills (1991, p. 37).

INSIDE THE NUCLEUS, DNA coils and uncoils, writhes and wriggles. Scientists often compare the form and movements of this long molecule to those of a snake. Molecular biologist Christopher Wills writes: "The two chains of DNA resemble two snakes coiled around each other in some elaborate courtship ritual."[11]

To sum up, DNA is a snake-shaped master of transformation that lives in water and is both extremely long and small, single and double.

Just like the cosmic serpent.

I KNEW THAT many shamanic peoples use images other than a "cosmic serpent" to discuss the creation of life, talking particularly of a rope, a vine, a ladder, or a stairway of celestial origin that links heaven and earth.

Mircea Eliade has shown that these different images form a common theme that he called the *axis mundi,* or axis of the world, and that he found in shamanic traditions the world over. According to Eliade, the axis mundi gives access to the Otherworld and to shamanic knowledge; there is a "paradoxical passage," normally reserved for the dead, that shamans manage to use while living, and this passage is often guarded by a serpent or a dragon. For Eliade, shamanism is the set of techniques that allows one to negotiate this passage, reach the axis, acquire the knowledge associated with it, and bring it back—most often to heal people.[12]

Here, too, the connection with DNA is clear. In the literature of molecular biology, DNA's shape is compared not only to two entwined serpents, but also, very precisely, to a rope, a vine, a ladder, or a stairway—the images varying from one author to another. For instance, Maxim Frank-Kamenetskii considers that "in a DNA molecule the complementary strands twine around one another like two lianas." Furthermore, scientists have recently begun to realize that many illnesses, including cancer, originate, and therefore may be solved, at the level of DNA.[13]

I set about exploring the different representations of the axis of the world, those images parallel to the cosmic serpent. The notion of an axis mundi is particularly common among the indigenous peoples of the Amazon. The Ashaninca, for example, talk of a "sky-rope." As Gerald Weiss writes: "Among the Campas there is a belief that at one time Earth and Sky were close together and connected by a cable. A vine called *inkíteca* (literally, "sky-rope"), with a peculiar stepped shape, was pointed out to the author as the cable that held Earth and Sky together."[14] According to Weiss, this vine is the same as the one indicated at the beginning of the twentieth century by the Taulipang Indians to Theodor Koch-Grünberg, one of the first ethnographers. In his work, Koch-Grünberg provided a skillful sketch of the Taulipang's vine.

*"Liana (*Bauhinia caulotretus*) 'that goes from earth up to heaven.'" From Koch-Grünberg (1917, Vol. 2, Drawing IV).*

Strangely, the Taulipang live in Guyana, some three thousand miles from the Ashaninca, yet they associate exactly the same vine with the sky-rope.

One of the best-known variants of the axis mundi is the caduceus, formed by two snakes wrapped around an axis. Since the most ancient times, one finds this symbol connected to the art of healing, from India to the Mediterranean. The Taoists of China represent the caduceus with the yin-yang, which symbolizes the coiling of two serpentine and complementary forms into a single androgynous vital principle[15]:

In the Western world, the caduceus continues to be used as the symbol of medicine, sometimes in modified form[16]:

Among the Shipibo-Conibo in the Peruvian Amazon, the axis mundi can be represented as a ladder. In the following drawing based on descriptions by ayahuasquero José Chucano Santos, the "sky-ladder" is surrounded by the cosmic anaconda Ronín (*see top of page 96*).

The ladder that gives access to shamanic knowledge is such a widespread theme that Alfred Métraux calls it the "symbol of the profession." He also reports that, as far as Amazonian shamans

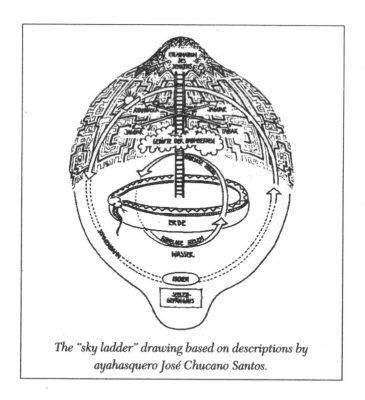

*The "sky ladder" drawing based on descriptions by
ayahasquero José Chucano Santos.*

are concerned, it is by contacting the "spirits of the ladder or of
the rungs" that they learn to "master all the secrets of magic."

Métraux also points out that these shamans drink "an infusion
prepared from a vine, the form of which suggests a ladder."[17] Indeed, the ayahuasca vine is often compared to a ladder, or even to
a *double helix,* as this photo taken by ethnobotanist Richard
Evans Schultes indicates (*see top of page 97*).

MOST OF THE CONNECTIONS I HAD FOUND up to this point between
the cosmic serpent and the axis of the world, and DNA, were related to *form.* This concurred with what Carlos Perez Shuma had
told me: Nature talks in *signs* and, to understand its language, one

"Banisteriopsis caapi, *a liana that tends to grow in charming double helices, is one of the primary ingredients in an entheogenic [hallucinogenic] potion known as . . . ayahuasca, yagé, caapi. . . . Those who know it call it 'spirit vine' or 'ladder to the Milky Way.' It is known also as ayahuasca ['vine of the soul'].*" (*Howard Rheingold quote.*) *From Schultes and Raffauf (1992, p. 26).*

has to pay attention to similarities in form. He had also said that the spirits of nature communicate with human beings in hallucinations and dreams—in other words, in mental *images*. This idea is common in "pre-rational" traditions. For instance, Heraclitus said of the Pythian oracle (from the Greek *puthôn*, "serpent") that it "neither declares nor conceals, but gives a sign."[18]

I wanted to go further than mere formal connections, however, and I knew, thanks to the work of Mircea Eliade, that shamans almost everywhere speak a secret language, "the language of all nature," which allows them to communicate with the spirits. I started looking for information about this phenomenon to see if there were any common elements in *content* between the language of the spirits of nature that shamans learn and the language of DNA.

Unfortunately, there are not many in-depth studies of shamanic language, no doubt because anthropologists have never really taken it seriously.[19] I found an exception in Graham Townsley's recent work on the songs of Yaminahua ayahuasqueros in the Peruvian Amazon.

According to Townsley, Yaminahua shamans learn songs, called *koshuiti*, by imitating the spirits they see in their hallucinations, in order to communicate with them. The words of these songs are almost totally incomprehensible to those Yaminahua who are not shamans. Townsley writes: "Almost nothing in these songs is referred to by its normal name. The abstrusest metaphoric circumlocutions are used instead. For example, night becomes 'swift tapirs,' the forest becomes 'cultivated peanuts,' fish are 'peccaries,' jaguars are 'baskets,' anacondas are 'hammocks' and so forth."

In each case, writes Townsley, the metaphorical logic can be explained by an obscure, but real, connection: "Thus fish become 'white-collared peccaries' because of the resemblance of a fish's gill to the white dashes on this type of peccary's neck; jaguars become 'baskets' because the fibers of this particular type of loose-woven basket (*wonati*) form a pattern precisely similar to a jaguar's markings."

The shamans themselves understand very clearly the meaning of these metaphors and they call them *tsai yoshtoyoshto*, literally "language-twisting-twisting." Townsley translates this expression as "twisted language."

The word *twist* has the same root as *two* and *twin*. *Twisted* means, technically, "double and wrapped around itself."

Why do Yaminahua shamans talk in twisted language? According to one of them: "With my koshuiti I want to see—singing, I carefully examine things—twisted language brings me close but not too close—with normal words I would crash into things— with twisted ones I circle around them—I can see them clearly."

For Townsley, all shamanic relations with the spirits are "deliberately constructed in an elliptical and multi-referential fashion so as to mirror the refractory nature of the beings who are their objects." He concludes: "*Yoshi* are real beings who are both 'like and not like' the things they animate. They have no stable or unitary nature and thus, paradoxically, the 'seeing as' of 'twisted language' is the only way of adequately describing them. Metaphor here is not improper naming but the only proper naming possible."[20]

I WENT ON to look for the connection between the language of spirits described by Yaminahua ayahuasqueros and the language of DNA. I found that "double and entwined," or "twisting-twisting," or "yoshtoyoshto," corresponded perfectly to the latter.

The genetic information of a human being (for example), called "genome," is contained in 3 billion letters spread out along a single filament of DNA. In some places, this filament winds around itself to form 23 more compact segments known as "chromosomes." We all inherit a complete set of chromosomes from each of our parents, and so we have 23 pairs of chromosomes. Each chromosome is made up of a very long thread of DNA which is already a double message to begin with—the main text on one ribbon, and the complementary duplicate on the other. Thus our cells all contain two complete genomes as well as their backup copies. Our genetic message is doubly double and contains a total of 6 billion base pairs, or 12 billion letters.

The DNA contained in the nucleus of a human cell is two

yards long, and the two ribbons that make up this filament wrap around each other several hundred million times.[21]

As far as its material aspect or its form is concerned, DNA is a doubly double text that wraps around itself. In other words, it is a "language-twisting-twisting."

THE TRANSCRIPTION ENZYMES read only the parts of the DNA text that code for the construction of proteins and enzymes. These passages, called "genes," represent only 3 percent of the human genome, according to various estimates. The remaining 97 percent are not read; their function is unknown.

Scientists have found spread out among the non-coding parts of the text a great number of endlessly repeated sequences with no apparent meaning, and even palindromes, which are words or sentences that can be read in either direction. They have called this apparent gibberish, which constitutes the overwhelming majority of the genome, "junk DNA."[22]

In this "junk," one finds tens of thousands of passages like this: ACACACACACACACACACACACACACACACAC. . . . There is even a 300-letter sequence that is repeated a total of half a million times. All told, repeat sequences make up a full third of the genome. Their meaning, so far, is unknown.

Molecular biologists Chris Calladine and Horace Drew sum up the situation: "The vast majority of DNA in our bodies does things that we do not presently understand."[23]

Scattered among this ocean of nonsense, genes are like islands where the language of DNA becomes comprehensible. Genes spell out the instructions for lining up amino acids into proteins. They do this with words of three letters. "CAG," for example, codes for amino acid glutamine in DNA language.

As all the words of the genetic code have three letters, and as

DNA has a four-letter alphabet (A,G,C,T), the genetic code contains $4 \times 4 \times 4 = 64$ possible words. These words all have a meaning and correspond either to one of the 20 amino acids used in the construction of proteins or to one of two punctuation marks ("start," "stop"). So there are 22 possible meanings for 64 words. This redundancy has led scientists to say that the genetic code is "degenerate." In fact, it simply has a wealth of synonyms—like a language where words as different as "jaguar" and "basket" have the same meaning.[24]

In reality, things are even more complex. Within genes, there are many non-coding segments called "introns." As soon as the transcription enzymes have transcribed a given gene, editing enzymes eliminate the introns with atomic precision and splice together the true coding segments, known as "exons." Some genes consist of up to 98 percent introns—which means that they contain only 2 percent genetic information. The role of these introns remains mysterious.[25]

The proportion of introns and exons in the human genome is not yet known, because so far, only half of all the genes it contains have been identified, out of a total estimated at 100,000.[26]

Along the DNA filament, "junk" and genes alternate; within genes, introns intermix with exons, which are themselves expressed in a language where almost every word has a synonym.

As far as both its content and its form are concerned, DNA is a doubly double language that wraps around itself.

Just like the twisted language of the spirits of nature.

WHAT DO THESE CONNECTIONS between DNA and the cosmic serpent, the axis of the world, and the language of the spirits of nature, mean?

The correspondences are too numerous to be explained by

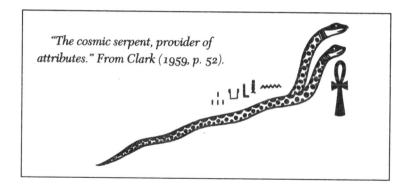

"The cosmic serpent, provider of attributes." From Clark (1959, p. 52).

chance alone. If I were a member of a jury having to pronounce itself on the matter, I would have the conviction that the same reality is being described from different perspectives.

Take the cosmic serpent of the Ancient Egyptians, the "provider of attributes." The signs that accompany it mean "one" (⏐), "several" (⏐⏐⏐), "spirit, double, vital force" (⊔), "place" (∟), "wick of twisted flax" (𝄁), and "water" (∿). Under the chin of the second serpent, there is an Egyptian cross meaning "key of life."[27]

The connections with DNA are obvious and work on all levels: DNA is indeed shaped like a long, single and double serpent, or a wick of twisted flax; it is a double vital force that develops from one to several; its place is water.

What else could the Ancient Egyptians have meant when they talked of a double serpent, provider of attributes and key of life, if not what scientists call "DNA"?

Why are these metaphors so consistently and so frequently used unless they mean what they say?

Chapter 8

THROUGH THE EYES
OF AN ANT

One sunny afternoon that spring I was sitting in the garden with my children. Birds were singing in the trees, and my mind began to wander. There I was, a product of twentieth-century rationality, my faith requiring numbers and molecules rather than myths. Yet I was now confronted with mythological numbers relative to a molecule, in which I had to believe. Inside my body sitting there in the garden sun were 125 billion miles of DNA. I was wired to the hilt with DNA threads and until recently had known nothing about it. Was this astronomical number really just a "useless but amusing fact,"[1] as some scientists would have it? Or did it indicate that the dimensions, at least, of our DNA are cosmic?

Some biologists describe DNA as an "ancient high biotechnology," containing "over a hundred trillion times as much information by volume as our most sophisticated information storage devices." Could one still speak of a technology in these circumstances? Yes, because there is no other word to qualify this duplicable, information-storing molecule. DNA is only ten atoms wide and as such constitutes a sort of ultimate technology: It is organic and so miniaturized that it approaches the limits of material existence.[2]

Shamans, meanwhile, claim that the vital principle that animates all living creatures comes from the cosmos and is *minded*. As ayahuasquero Pablo Amaringo says: "A plant may not talk, but there is a spirit in it that is conscious, that sees everything, which is the soul of the plant, its essence, what makes it alive." According to Amaringo, these spirits are veritable beings, and humans are also filled with them: "Even the hair, the eyes, the ears are full of beings. You see all this when ayahuasca is strong."[3]

During the past weeks, I had come to consider that the perspective of biologists could be reconciled with that of ayahuasqueros and that both could be true at the same time. According to the stereoscopic image I could see by gazing at both perspectives simultaneously, DNA and the cell-based life it codes for are an extremely sophisticated technology that far surpasses our present-day understanding and that was initially developed elsewhere than on earth—which it radically transformed on its arrival some four billion years ago.

This point of view was completely new to me and had changed my way of looking at the world. For instance, the leaves of trees now appeared to be true solar panels. One had only to look at them closely to see their "technological," or organized, aspect (*see top of page 105*).

This revelation was troubling. I started thinking about my eyes, through which I was looking at the plants in the garden. Over the course of my readings, I had learned that the human eye is more sophisticated than any camera of similar size. The cells on the outer layer of the retina can absorb a single particle of light, or photon, and amplify its energy at least a million times, before transferring it in the form of a nervous signal to the back of the

A magnified section of a leaf illustrating its organized,
technological aspect.

brain. The iris, which functions as the eye's diaphragm, is automatically controlled. The cornea has just the right curvature. The lens is focused by miniature muscles, which are also controlled automatically by feedback. The final result of this visual system, still imperfectly understood in its entirety, is a clear, colored, and three-dimensional image inside the brain that we perceive as external. We never see reality, but only an internal representation of it that our brain constructs for us continuously.[4]

What troubled me was not so much the resemblance of the human eye to an organic and extremely sophisticated technology born of cosmic knowledge, but that they were *my* eyes. Who was this "I" perceiving the images flooding into my mind? One thing was sure: I was not responsible for the construction of the visual system with which I was endowed.

I did not know what to make of these thoughts. Staring blankly at the lawn in front of me, I started following a shiny, black ant making its way across the thick blades of grass with the determination of a tank. It was heading toward the colony of aphids in the tree at the bottom of the garden. This was an ant belonging to a species that herds aphids and "milks" them for their sweet secretions.

I began thinking that this ant had a visual system quite different from my own that apparently functioned every bit as well. Despite our differences in size and shape, our genetic information was written in the same language—which we were both incapable of seeing, given that DNA is smaller than visible light, even to the eyes of an ant.

I found it interesting that the language containing the instructions for the creation of different visual systems should be itself invisible. It was as if the instructions were to remain hidden from their beneficiaries, as if we were wired in such a way that we could not see the wires. . . .

Why?

I tried reconsidering the question from a "shamanic" point of view. It was as if these beings inside us wanted to hide. . . . *But that's what the Ashaninca say! They call the invisible beings who created life the "maninkari," literally "those who are hidden"!*

LATER THAT AFTERNOON, I returned to my office and started rereading the passages concerning the maninkari in Gerald Weiss's exhaustive study on Ashaninca cosmology. According to Weiss, the Ashaninca believe that the most powerful of all maninkari is the "Great Transformer" Avíreri, who created life on earth, starting with the seasons and then moving on to the entirety of living beings. Accompanied sometimes by his sister, at others by his nephew, Avíreri is one of the divine trickster twins who create by transformation and are so common in mythology.

It was in reading the last story about the end of Avíreri's trajectory that I had a shock. Having completed his creation work, Avíreri goes to a party where he gets drunk on manioc beer. His sister, who is also a trickster, invites him to dance and pushes him into a hole dug in advance. She then pretends to pull him up by throwing him a *thread,* then a *cord*—but neither is strong enough. Furious with his sister, whom he transforms into a tree, Avíreri decides to escape by digging a hole into the *underworld.* He ends up at a place called *River's End,* where a *strangler vine* wraps around him. From there, he continues to sustain to this day *his numerous children on earth.*[5]

How could I have missed the connections between the twin being Avíreri, the Great Transformer, and the DNA double helix, first creating the breathable atmosphere ("the seasons"), then the entirety of living beings by transformation, living in the microscopic world ("underworld"), in cells filled with seawater ("River's End"), taking the form of a thread, a cord, or a strangler vine which wraps around itself, and, finally, sustaining to this day all the living species of the planet?

For weeks I had been finding connections between myths and molecular biology. I was not even surprised to see that the creation myth of an indigenous Amazonian people coincided with the description made by today's biologists of the development of life on earth. What shook me, and even filled me with consternation, was that I had had this evidence under my nose for years without giving it the slightest importance. My gaze had been too narrow.

Sitting in my office, I remembered the time Carlos Perez Shuma had told me, "The maninkari taught us how to spin and weave cotton." Now the meaning seemed obvious; the two ribbons of the DNA double helix wrap around each other 600 million times inside each human cell: "Who else could have taught us to weave?" The problem for me was that I had not believed him. I had not considered for one moment that his words corresponded to something real.

Under these circumstances, what did my title "doctor of anthropology" signify—other than an intellectual imposture in relation to my object of study?

These revelations overwhelmed me. To make amends, I resolved then and there to take shamans at their word for the rest of my investigation.

WHAT HAD BECOME of the investigation that posed the enigma of the hallucinatory knowledge of Western Amazonia's indigenous people? Why had it ended up with cosmic serpents from around the world entwined with DNA molecules?

For some weeks now, I had been in a sort of trance, my mind flooded with an almost permanent flux of strange, if not impossible, connections. My only discipline had been to note them down, or to tape them, instead of repressing them out of disbelief. My worldview had been turned upside down, but I was slowly coming back to my senses, and the first question I asked myself was: What did all this mean?

I was now of the opinion that DNA was at the origin of shamanic knowledge. By "shamanism," I understood a series of defocalization techniques: controlled dreams, prolonged fasting, isolation in wilderness, ingestion of hallucinogenic plants, hypnosis based on a repetitive drumbeat, near-death experience, or a combination of the above. Aboriginal shamans of Australia reach conclusions similar to those of Amazonian ayahuasqueros, without the use of psychoactive plants, by working mainly with their dreams. What techniques did Chuang-Tzu, the Egyptian pharaohs, and the animists of Benin use, to name but a few? Who could say? But they all spoke, in one form or another, of a cosmic serpent— as did the Australians, the Amazonians, and the Aztecs.

By using these different techniques, it therefore seemed possible to induce neurological changes that allow one to pick up information from DNA. But from which DNA? At first I thought that I had found the answer when I learned that, in each human cell, there is the equivalent of "the information contained in one thousand five hundred encyclopedia volumes"[6]—in other words, the equivalent of a bookcase about ten yards long and two yards high. There, I thought, is the origin of knowledge.

On reflection, however, I saw that this idea was improbable. There was no reason why the human genome, no matter how vast, should contain information about the Amazonian plants necessary for the preparation of curare, for example. Furthermore, the ayahuasqueros said that the highly sophisticated sound-images that they saw and heard in their hallucinations were *interactive,* and that it was possible to communicate with them. These images could not originate from a static, or textual, set of information such as 1,500 encyclopedia volumes.

My own experience with ayahuasca-induced hallucinations was limited, but was sufficient to suggest a trail. *Ayahuasquero* Ruperto Gomez, who had initiated me, had called the hallucinogenic brew "the television of the forest," and I had indeed seen sequences of hallucinatory images flashing by at blinding speed, as if they were truly transmitted from outside my body, but picked up inside my head.[7]

I knew of no neurological mechanism on which to base this working hypothesis, but I did know that DNA was an aperiodic crystal that traps and transports electrons with efficiency and that emits photons (in other words, electromagnetic waves) at ultraweak levels currently at the limits of measurement—and all this more than any other living matter.[8] This led me to a potential candidate for the transmissions: the global network of DNA-based life.

All living beings contain DNA, be they bacteria, carrots, or humans. DNA, as a substance, does not vary from one species to another; only the order of its letters changes. This is why biotechnology is possible. For instance, one can extract the DNA sequence in the human genome containing the instructions to build the insulin protein and splice it into the DNA of a bacterium, which will then produce insulin similar to that normally excreted by the human pancreas. The cellular machines called ribosomes, which assemble the proteins inside the bacterium, understand the same four-letter language as the ribosomes inside human pancreatic cells and use the same 20 amino acids as building blocks. Biotechnology proves by its very existence the fundamental unity of life.

Each living being is constructed on the basis of the instructions written in the informational substance that is DNA. A single bacterium contains approximately ten million units of genetic information, whereas a microscopic fungus contains a billion units. In a mere handful of soil, there are approximately ten billion bacteria and one million fungi. This means that there is more order, and information, in a handful of earth than there is on the surfaces of all the other known planets combined.[9] The information contained in DNA makes the difference between life and inert matter.

The earth is surrounded by a layer of DNA-based life that made the atmosphere breathable and created the ozone layer, which protects our genetic matter against ultraviolet and mutagenic rays. There are even anaerobic bacteria living half a mile beneath the ocean floor; the planet is wired with life deep into its crust.[10]

When we walk in a field, DNA and the cell-based life it codes for are everywhere: inside our own bodies, but also in the puddles, the mud, the cow pies, the grass on which we walk, the air we breathe, the birds, the trees, and everything that lives.

This global network of DNA-based life, this biosphere, encircles the entire earth.

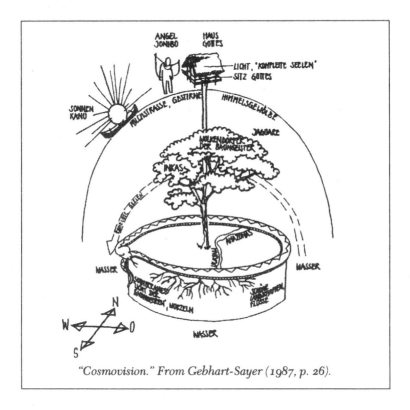

"Cosmovision." From Gebhart-Sayer (1987, p. 26).

What better image for the DNA-based biosphere than Ronín, the cosmic anaconda of the Shipibo-Conibo? The anaconda is an amphibious snake, capable of living both in water and on land, just like the biosphere's creatures. Ayahuasquero Laureano Ancon explains the above image: "The earth upon which we find ourselves is a disk floating in great waters. The serpent of the world Ronín is half-submerged and surrounds it entirely."[11]

Here is, according to my conclusions, the great instigator of the hallucinatory images perceived by ayahuasqueros: the crystalline and biospheric network of DNA-based life, alias the cosmic serpent.

DURING MY FIRST AYAHUASCA EXPERIENCE I saw a pair of enormous and terrifying snakes. They conveyed an idea that bowled me over and later encouraged me to reconsider my self-image. They taught me that I was just a human being. To others, this may not seem like a great revelation; but at the time, it was exactly what the young anthropologist I was needed to learn. Above all, it was a thought that I could not have had by myself, precisely because of my anthropocentric presuppositions.

I also felt very clearly that the speed and the coherence of certain sequences of images could not have come from the chaotic storage room of my memory. For example, I saw in a dizzying visual parade the superimposing of the veins of a human hand on those of a green leaf. The message was crystal clear: We are made of the same fabric as the vegetal world. I had never really thought of this so concretely. The day after the ayahuasca session, I felt like a new being, united with nature, proud to be human and to belong to the grandiose web of life surrounding the planet. Once again, this was a totally new and constructive perspective for the materialistic humanist that I was.

This experience troubled me deeply. If I was not the source of these highly coherent and educational images, where did they come from? And who were those snakes who seemed to know me better than myself? When I asked Carlos Perez Shuma, his answer was elliptic: All I had to do was take the snakes' picture the

Detail from Pablo Amaringo's painting "Pregnant by an Anaconda," reproduced in Luna and Amaringo (1991, p. 111).

next time I saw them. He did not deny their existence—on the contrary, he implied that they were as real as the reality we are all familiar with, if not more so.

Eight years after my first ayahuasca experience, my desire to understand the mystery of the hallucinatory serpents was undiminished. I launched into this investigation and familiarized myself with the different studies of ayahuasca shamanism only to discover that my experience had been commonplace. People who drink ayahuasca see colorful and gigantic snakes more than any other vision[12]—be it a Tukano Indian, an urbanized shaman, an anthropologist, or a wandering American poet.[13] For instance, serpents are omnipresent in the visionary paintings of Pablo Amaringo[14] (*see above*).

Over the course of my readings, I discovered that the serpent was associated just about everywhere with shamanic knowledge—even in regions where hallucinogens are not used and where snakes are unknown in the local environment. Mircea Eliade says that in Siberia the serpent occurs in shamanic ideology and in the shaman's costume among peoples where "the reptile itself is unknown."[15]

Then I learned that in an endless number of myths, a gigantic and terrifying serpent, or a dragon, guards the axis of knowledge, which is represented in the form of a ladder (or a vine, a cord, a tree . . .). I also learned that (cosmic) serpents abound in the creation myths of the world and that they are not only at the origin of knowledge, but of life itself.

Snakes are omnipresent not only in the hallucinations, myths, and symbols of human beings in general, but also in their dreams. According to some studies, "Manhattanites dream of them with the same frequency as Zulus." One of the best-known dreams of this sort is August Kekulé's, the German chemist who discovered the cyclical structure of benzene one night in 1862, when he fell asleep in front of the fire and dreamed of a snake dancing in front of his eyes while biting its tail and taunting him. According to one commentator, "There is hardly any need to recall that this contribution was fundamental for the development of organic chemistry."[16]

Why do life-creating, knowledge-imparting snakes appear in the visions, myths, and dreams of human beings around the world?

The question has been asked, and a simple and neurological answer has been proposed and generally accepted: because of the instinctive fear of venom programmed into the brains of primates such as ourselves. Balaji Mundkur, author of the only global study on the matter, writes, "The fundamental cause of the origin of serpent cults seems to be unlike any which gave rise to practically all other animal cults; that fascination by, and awe of, the serpent appears to have been compelled not only by elementary fear of its venom, but also by less palpable, though quite primordial psychological sensitivities rooted in the evolution of the primates; that unlike almost all other animals, serpents, in varying degree, provoke certain characteristically intuitive, irrational, phobic responses in human and nonhuman primates alike; . . . and that the serpent's power to fascinate certain primates is dependent on the reaction of the latter's autonomic nervous system to the mere sight of reptilian sinuous movement—a type of response that may have been reinforced by memories of venomous attacks during anthropogenesis and the differentiation of human societies. . . . The fascination of serpents, in short, is synonymous with a state of fear that amounts, at least temporarily, to *morbid revulsion* or phobia . . . whose symptoms few other species of animals—perhaps none—can elicit" (original italics).[17]

In my opinion, this is a typical example of a reductionist, illogical, and inexact answer. Do people really venerate what they fear most? Do people suffering from phobia of spiders, for instance, decorate their clothes with images of spiders, saying, "We venerate these animals because we find them repulsive"? Hardly. Therefore, I doubt that Siberian shamans embellish their costumes with a great number of *ribbons representing serpents* simply because they suffer from a phobia of these reptiles. Besides, most of the serpents found in the costumes of Siberian shamans do not represent real animals, but snakes with two tails. In a great number of creation myths, the serpent that plays the main part is not a real reptile; it is a cosmic serpent and often has two heads, two feet, or two wings or is so big that it wraps around the earth. Furthermore, *venerated serpents are often nonvenomous*. In the Amazon, the nonvenomous snakes such as anacondas and boas are the ones that people consider sacred, like the cosmic anaconda Ronín. There is no lack of aggressive and deadly snakes with devastating venom in the Amazon, such as the bushmaster and the fer-de-lance, which are an everyday threat to life—and yet, they are never worshipped.[18]

The answer, for me, lies elsewhere—which does not mean that primates do not suffer from an instinctive, or even a "programmed," fear of snakes. My answer is speculative, but could not be more restricted than the generally accepted theory of venom phobia. It is that the global network of DNA-based life emits ultra-weak radio waves, which are currently at the limits of measurement, but which we can nonetheless perceive in states of defocalization, such as hallucinations and dreams. As the aperiodic crystal of DNA is shaped like two entwined serpents, two ribbons, a twisted ladder, a cord, or a vine, we see in our trances serpents, ladders, cords, vines, trees, spirals, crystals, and so on. Because DNA is a master of transformation, we also see jaguars, caymans, bulls, or any other living being. But the favorite newscasters on DNA-TV seem unquestionably to be enormous, fluorescent serpents.

This leads me to suspect that the cosmic serpent is narcissistic—or, at least, obsessed with its own reproduction, even in imagery.

RECEPTORS AND
TRANSMITTERS

M y investigation had led me to formulate the following working hypothesis: In their visions, shamans take their consciousness down to the molecular level and gain access to information related to DNA, which they call "animate essences" or "spirits." This is where they see double helixes, twisted ladders, and chromosome shapes. This is how shamanic cultures have known for millennia that the vital principle is the same for all living beings and is shaped like two entwined serpents (or a vine, a rope, a ladder . . .). DNA is the source of their astonishing botanical and medicinal knowledge, which can be attained only in defocalized and "nonrational" states of consciousness, though its results are empirically verifiable. The myths of these cultures are filled with biological imagery. And the shamans' metaphoric explanations correspond quite precisely to the descriptions that biologists are starting to provide.

I knew this hypothesis would be more solid if it rested on a neurological basis, which was not yet the case. I decided to direct my investigation by taking ayahuasqueros at their word—and they unanimously claimed that certain psychoactive substances (containing molecules that are active in the human brain) influence the spirits in precise ways. The Ashaninca say that by ingesting ayahuasca or tobacco, it is possible to see the normally invisible and hidden maninkari spirits. Carlos Perez Shuma had told me that tobacco attracted the maninkari. Amazonian shamans in general consider tobacco a food for the spirits, who crave it "since they no longer possess fire as human beings do."[1] If my hypothesis were correct, it ought to be possible to find correspondences between these shamanic notions and the facts established by the study of the neurological activity of these same substances. More precisely, there ought to be an analogous connection between nicotine and DNA contained in the nerve cells of a human brain.

The idea that the maninkari liked tobacco had always seemed funny to me. I considered "spirits" to be imaginary characters who could not really enjoy material substances. I also considered smoking to be a bad habit, and it seemed improbable that spirits (inasmuch as they existed) would suffer from the same kinds of addictive behaviors as human beings. Nevertheless, I had resolved to stop letting myself be held up by such doubts and to pay attention to the literal meaning of the shamans' words, and the shamans were categorical in saying that spirits had an almost insatiable hunger for tobacco.[2]

I started following this trail by spending a few days at the library. I even made several phone calls to a specialist in the neurological mechanisms of nicotine to deepen my understanding and make sure I was not establishing imaginary connections—neurology being the last of my competencies. Here is what I learned.

In the human brain, each nerve cell, or neuron, has billions of receptors on its outer surface. These receptors are proteins specialized in the recognition and trapping of specific neurotransmitters, or similar substances. A molecule of nicotine shares structural similarities with the neurotransmitter acetylcholine and fits like a skeleton key into its receptor on certain neurons.[3] This receptor is embedded in the cell's membrane and is a large protein that includes not only a "lock" (the docking site for the external molecule), but also a channel, with a gate that is normally shut. When a key is introduced into the lock—when a molecule of nicotine fits into the binding site at the top of the receptor—the channel's gate opens, allowing in a selective flow of positively charged atoms of calcium and sodium. The latter trigger a (poorly understood) cascade of electric reactions inside the cell, which ends up exciting the DNA contained in the nucleus, causing it to activate several genes, including those corresponding to the proteins that make up nicotinic receptors.[4]

The more you give nicotine to your neurons, the more the DNA they contain activates the construction of nicotinic receptors, within certain limits. Here, I thought, is the almost insatiable hunger of the spirits for tobacco: The more you give them, the more they want.

I was surprised by the degree of correspondence between shamanic notions of tobacco and neurological studies of nicotine. One only had to do a literal translation to pass from one to the other. However, scientific accounts in terms of "receptors," "flux of positively charged atoms," and "stimulation of the transcription of the genes coding for the subunits of nicotinic receptors" did not explain in any way the effects of nicotine on consciousness. How was it that shamans *saw* spirits by ingesting staggering quantities of tobacco?

Before continuing with this question, I will clarify two points. First, the discovery that nicotine stimulates the construction of nicotinic receptors was only made at the beginning of the 1990s;

the connection between this phenomenon and the addiction displayed by tobacco users seems obvious, but has yet to be explored in detail.

Second, there are fundamental differences between the shamanic use of tobacco and the consumption of industrial cigarettes. The botanical variety used in the Amazon contains up to eighteen times *more* nicotine than the plants used in Virginia-type cigarettes. Amazonian tobacco is grown without chemical fertilizers or pesticides and contains none of the ingredients added to cigarettes, such as aluminum oxide, potassium nitrate, ammonium phosphate, polyvinyl acetate, and a hundred or so others, which make up approximately 10 percent of the smokable matter.[5] During combustion, a cigarette emits some 4,000 substances, most of which are toxic. Some of these substances are even radioactive, making cigarettes the largest single source of radiation in the daily life of an average smoker. According to one study, the average smoker absorbs the equivalent of the radiation dosages from 250 chest X-rays per year. Cigarette smoke is directly implicated in more than 25 serious illnesses, including 17 forms of cancer.[6] In the Amazon, on the other hand, tobacco is considered a *remedy*. The Ashaninca word for "healer," or "shaman," is *sheripiári*—literally, "the person who uses tobacco."[7] The oldest Ashaninca men I knew were all sheripiári. They were so old that they did not know their own age, which only their deeply wrinkled skin suggested, and they were remarkably alert and healthy.

Intrigued by these disparities, I looked through data banks for comparative studies between the toxicity of the Amazonian variety (*Nicotiana rustica*) and the variety used by the manufacturers of cigarettes, cigars, rolling tobacco, and pipe tobacco (*Nicotiana tabacum*). I found nothing. The question, it seemed, had not been asked. I also looked for studies on the cancer rate among shamans who use massive and regular doses of nicotine: again, nothing. So I decided to write to the main authority on the matter, Johannes Wilbert, author of the book *Tobacco and shamanism in South America,* to put my questions to him. He replied: "There is certainly evidence that Western tobacco products contain many different harmful agents which are probably not present in organically grown plants. I have not heard of shamans developing cancers but that may, of course, be a function of several things like lack of Western diagnosis, natural life span of indigenous people, magico-religious restriction of tobacco use in tribal societies, etc."[8]

It seems clear that nicotine does not cause cancer, given that it is active in the brain and that cigarettes do not cause cancer in the brain, but in the lungs, esophagus, stomach, pancreas, rectum, kidneys, and bladder, the organs reached by the *carcinogenic tars,* which are also swallowed.

In any case, scientists have never really considered tobacco as a hallucinogen, because Westerners have never smoked large enough doses to reach the hallucinatory state.[9] Consequently, the neurological mechanisms of hallucinations induced by tobacco have not been studied. Paradoxically, nicotinic receptors are the ones best known to neurologists, who have been studying them for decades, given that there are both substances that stimulate these receptors, like acetylcholine and nicotine, and others that block them, like curare and the venom of certain snakes.[10] Indeed, by one of those curious coincidences, tobacco, curare, and snake venom all fit into exactly the same locks inside our brains.

AS THE NEUROLOGICAL TRAIL of tobacco-induced hallucinations was a dead end, I turned to ayahuasca. Carlos Perez Shuma had said: "When an ayahuasquero drinks his plant mixture, the spirits present themselves to him and explain everything." The shamans of Western Amazonia in general claim that their hallucinogenic brew allows them to see the spirits. According to my hypothesis, there ought to be a demonstrable connection between the active ingredients of ayahuasca and the DNA contained in the nerve cells of a human brain. I went looking for it.

Ayahuasca is the most botanically and chemically complex hallucinogen. It can be thought of as a psychoactive cocktail, containing different additives depending on the region, the practitioner, and the desired effects. Scientists who have studied its composition agree that dimethyltryptamine is its main active ingredient. This highly hallucinogenic substance seems also to be produced in small quantities by the human brain. However, since the end of the 1960s, dimethyltryptamine has been at the top of the controlled substances list, along with synthetic compounds such as heroin and LSD. This means not only that it is illegal for the average person, but that scientific studies on its effects are discouraged, and rare.[11]

In the literature, I found only one scientific investigation on dimethyltryptamine that had been carried out under neutral conditions: For once, the hallucinogen was not considered as a "psychotomimetic" (that is, "imitator of psychosis"), its "psychopathology" was not discussed, and it was not administered to imprisoned criminals playing the part of human guinea pigs. In the 1994 study published by Rick Strassman and colleagues, the subjects were all experienced hallucinogen users who chose to participate in the research. With one exception, they were all professionals or students in professional training programs.[12]

The authors of this study devote a paragraph to the contents of their subjects' visions: images that "were both familiar and novel, such as 'a fantastic bird,' 'a tree of life and knowledge,' 'a ballroom with crystal chandeliers,' human and 'alien' figures (such as 'a little round creature with one big eye and one small eye, on nearly invisible feet'), 'the inside of a computer's boards,' 'ducts,' 'DNA double helices,' 'a pulsating diaphragm,' 'a spinning gold disc,' 'a huge fly eye bouncing in front of my face,' tunnels and stairways."[13]

Under the influence of dimethyltryptamine, people saw trees of life and knowledge, crystals, stairways, and DNA double helices. This confirmed my hypothesis that shamans perceive images containing biomolecular information—but in no way explained its mechanism. How was it that molecular reality became accessible to the normally nonmolecular consciousness of human beings? What went on in the brain for normal consciousness to disappear in a flood of strange images?

Knowledge about the neurological pathways of hallucinogenic substances has made great progress in recent years. While scientists have known for over a quarter of a century that molecules such as dimethyltryptamine, psilocybin, and even LSD resemble the neurotransmitter serotonin, it was only in the 1990s that they discovered the existence of seven types of serotonin receptors, in relation to which each hallucinogen has a specific mode of functioning.[14]

One of these receptors is built on the model of the lock-coupled-to-a-channel. The others are more like "antennae," which span the cell's membrane. When a molecule of serotonin stimulates the external part of the antenna, the latter sets off a signal inside the cell.[15]

I looked for a connection between the stimulation of serotonin receptors and DNA and found a recent (1994) article entitled

"Serotonin increases DNA synthesis in rat proximal and distal pulmonary vascular smooth muscle cells in culture." The connection existed, but was still not very clear, as the increase in DNA activity following an input of serotonin was measurable, but the cascade of reactions inside the cell, from the antenna to the nucleus, remained hypothetical.[16]

To my knowledge, current research on the neurological mechanisms of hallucinogens stops at these questions of receptors. Metaphorically speaking, we now understand where the electricity comes from and where the plug is, but we still do not know how the television works.

CURRENTLY, DNA is not part of the scientific discussion on hallucinations, but this has not always been the case. At the end of the 1960s, the uneasiness about the casual and large-scale use of LSD generated the rumor that hallucinogens "break chromosomes." In the ensuing hysteria, all kinds of poorly conceived experiments seemed to confirm this hypothesis. For instance, researchers administered the equivalent of more than three thousand LSD doses to female monkeys in their fourth month of pregnancy; at birth, one infant monkey was stillborn, two others showed "facial deformities," and a fourth died after a month—mainly proving that these animals had been severely and unnecessarily illtreated. Other researchers noticed that naked DNA, extracted from the cell's nucleus and placed in a test tube, attracted LSD and other hallucinogenic molecules; according to their calculations, these molecules intercalated between the rungs of the ladder formed by the double helix, thereby causing the famous "chromosome breaks."[17] (Later it was pointed out that naked DNA attracted thousands of substances in this way.)

Several scientists suggested, on the basis of this research, that DNA played a role in hallucinatory mechanisms.[18] However, this idea did not receive much attention in the charged atmosphere of the times. On the contrary, scientific research on these substances was abandoned during the first half of the 1970s.

In those days, the scientific understanding of DNA and cellular receptors was embryonic. Researchers did not know that DNA was never naked in biological reality, but was always wrapped up in proteins inside the nucleus, and that the latter was never penetrated by extracellular hallucinogenic molecules. It wasn't until the 1980s that scientists understood that hallucinogens stimulated receptors situated on the outside of cells.[19]

From the middle of the 1970s onward, the connection between DNA and hallucinogens disappears from the scientific literature.[20] It would no doubt be interesting to reconsider it in the light of the new knowledge established by molecular biology.

LIKE THE AXIS MUNDI of shamanic traditions, DNA has the form of a twisted ladder (or a vine . . .); according to my hypothesis, DNA was, like the axis mundi, the source of shamanic knowledge and visions. To be sure of this I needed to understand how DNA could transmit visual information. I knew that it emitted photons, which are electromagnetic waves, and I remembered what Carlos Perez Shuma had told me when he compared the spirits to "radio waves" ("Once you turn on the radio, you can pick them up. It's like that with souls; with ayahuasca and tobacco, you can see them and hear them"). So I looked into the literature on photons of biological origin, or "biophotons."

In the early 1980s, thanks to the development of a sophisticated measurement device, a team of scientists demonstrated that the cells of all living beings emit photons at a rate of up to approximately 100 units per second and per square centimeter of surface area. They also showed that DNA was the source of this photon emission.[21]

During my readings, I learned with astonishment that the wavelength at which DNA emits these photons corresponds exactly to the narrow band of *visible light:* "Its spectral distribution ranges at least from infrared (at about 900 nanometers) to ultraviolet (up to about 200 nanometers)."[22]

This was a serious trail, but I did not know how to follow it. There was no proof that the light emitted by DNA was what shamans saw in their visions. Furthermore, there was a fundamental aspect of this photon emission that I could not grasp. According to the researchers who measured it, its weakness is such that it corresponds "to the intensity of a candle at a distance of about 10 kilometers," but it has "a surprisingly high degree of coherence, as compared to that of technical fields (laser)."[23] How could an ultra-weak signal be highly coherent? How could a distant candle be compared to a "laser"?

After thinking about it at length, I came to understand that the coherence of biophotons depended not so much on the intensity of their output as on its regularity. In a coherent source of light, the quantity of photons emitted may vary, but the emission intervals remain constant.

DNA emits photons with such regularity that researchers compare the phenomenon to an "ultra-weak laser." I could understand that much, but still could not see what it implied for my investigation. I turned to my scientific journalist friend, who explained it immediately: "A coherent source of light, like a laser, gives the sensation of bright colors, a luminescence, and an impression of holographic depth."[24]

My friend's explanation provided me with an essential element. The detailed descriptions of ayahuasca-based hallucinatory experiences invariably mention bright color, and, according to the authors of the dimethyltryptamine study: "Subjects described the colors as brighter, more intense, and deeply saturated than those seen in normal awareness or dreams: 'It was like the blue of a desert sky, but on another planet. The colors were 10 to 100 times more saturated.'"[25]

It was almost too good to be true. DNA's highly coherent photon emission accounted for the luminescence of hallucinatory images, as well as their three-dimensional, or holographic, aspect.

On the basis of this connection, I could now conceive of a neurological mechanism for my hypothesis. The molecules of nicotine or dimethyltryptamine, contained in tobacco or ayahuasca, activate their respective receptors, which set off a cascade of electrochemical reactions inside the neurons, leading to the stimulation of DNA and, more particularly, to its emission of visible waves, which shamans perceive as "hallucinations."[26]

There, I thought, is the source of knowledge: DNA, living in water and emitting photons, like an aquatic dragon spitting fire.

IF MY HYPOTHESIS IS CORRECT, and if ayahuasqueros perceive DNA-emitted photons in their visions, it ought to be possible to find a link between these photons and consciousness. I started looking for it in the biophoton literature.

Researchers working in this new field mainly consider biophoton emission as a "cellular language" or a form of "nonsubstantial biocommunication between cells and organisms." Over the last fifteen years, they have conducted enough reproducible experiments to believe that cells use these waves to direct their own internal reactions as well as to communicate among themselves, and even between organisms. For instance, photon emission provides a communication mechanism that could explain how billions of individual plankton organisms cooperate in swarms, behaving like "super-organisms."[27]

Biophoton emission may fill certain gaps in the theories of orthodox biology, which center exclusively on molecules. Yet researchers in this new field of inquiry will have to work hard to convince the majority of their colleagues. As Mae-Wan Ho and Fritz-Albert Popp point out, many biologists find the idea that the cell is a solid-state system difficult to imagine, "as few of us have the requisite biophysical background to appreciate the implications."[28] But this did not help my search for a connection between DNA-emitted photons and consciousness. I did not find a publication dealing with this connection or, for that matter, with the subject of the influence of nicotine or dimethyltryptamine on biophoton emission.

So I decided to call Fritz-Albert Popp in his university laboratory in Germany. He was kind enough to spare his time to an unknown anthropologist conducting an obscure investigation. During the conversation, he confirmed a good number of my impressions. I ended up asking him whether he had considered the possibility of a connection between DNA's photon emission and consciousness. He replied: "Yes, consciousness could be the electromagnetic field constituted by the sum of these emissions. But, as you know, our understanding of the neurological basis of consciousness is still very limited."[29]

ONE THING had struck me as I went over the biophoton literature. Almost all of the experiments conducted to measure biophotons involved the use of *quartz*. As early as 1923, Alexander Gurvich noticed that cells separated by a quartz screen mutually influenced each other's multiplication processes, which was not the case with a metal screen. He deduced that cells emit electromagnetic waves with which they communicate. It took more than half a century to develop a "photomultiplier" capable of measuring this ultra-weak radiation; the container of this device is also made of quartz.[30]

Quartz is a crystal, which means it has an extremely regular arrangement of atoms that vibrate at a very stable frequency. These characteristics make it an excellent receptor and emitter of electromagnetic waves, which is why quartz is abundantly used in radios, watches, and most electronic technologies.

Quartz crystals are also used in shamanism around the world. As Gerardo Reichel-Dolmatoff writes: "Quartz crystals, or translucent rock crystals, have played a major role in shamanic beliefs and practices at many times in history and in many parts of the world. They have frequently been found in prehistoric contexts; they are mentioned in many early sources; they were prominent in Old World alchemy, witchcraft, and magic, and they are still in use in many traditional societies. American Indian shamans and healers use rock crystals for curing, scrying, and many other purposes, and their ancient use in the Americas is known from archaeological reports."[31]

Amazonian shamans, in particular, consider that spirits can materialize and become visible in quartz crystals. Some sheripiári even feed tobacco juice to their stones daily.[32]

What if these spirits were none other than the biophotons emitted by all the cells of the world and were picked up, amplified, and transmitted by shamans' quartz crystals, Gurvich's quartz screens, and the quartz containers of biophoton researchers? This would mean that spirits are beings of pure light—as has always been claimed.

DNA IS ALSO A CRYSTAL, as molecular biologist Maxim Frank-Kamenetskii explains: "The base pairs in it are arranged as in a crystal. This is, however, a linear, one-dimensional crystal, with each base pair flanked by only two neighbors. The DNA crystal is aperiodic, since the sequence of base pairs is as irregular as the sequence of letters in a coherent printed text. . . . Thus, it came as no surprise that the one-dimensional DNA crystal, a crystal of an entirely new type, had very much intrigued physicists."[33]

The four DNA bases are hexagonal (like quartz crystals), but they each have a slightly different shape.[34] As they stack up on top of each other, forming the rungs of the twisted ladder, they line up in the order dictated by the genetic text. Therefore, the DNA double helix has a slightly irregular, or aperiodic, structure. However, this is not the case for the repeat sequences that make up a full third of the genome, such as ACACACACACACAC. In these sequences, DNA becomes a regular arrangement of atoms, a *periodic* crystal—which could, by analogy with quartz, pick up as many photons as it emits. The variation in the length of the repeat sequences (some of which contain up to 300 bases) would help pick up different frequencies and could thereby constitute a possible and new function for a part of "junk" DNA.[35]

I suggest this because my hypothesis requires a receptor as much as an emitter. For the moment, the reception of biophotons has not been studied.[36]

Even DNA's emission of photons remains mysterious, and no one has been able to establish its mechanism directly. Naked DNA, extracted from the cell's nucleus, emits photons so weakly as to escape measurement.[37]

Despite these uncertainties, I wish to develop my hypothesis further by proposing the following idea: What if DNA, stimulated by nicotine or dimethyltryptamine, activates not only its *emission* of photons (which inundate our consciousness in the form of hallucinations), but also its *capacity to pick up* the photons emitted by the global network of DNA-based life? This would mean that the biosphere itself, which can be considered "as a more or less fully interlinked unit,"[38] is the source of the images.

Chapter 10

BIOLOGY'S
BLIND SPOT

I began my investigation with the enigma of "plant communication." I went on to accept the idea that hallucinations could be
a source of verifiable information. And I ended up with a hypothesis suggesting that a human mind can communicate in defocalized consciousness with the global network of DNA-based life.
All this contradicts principles of Western knowledge.

Nevertheless, my hypothesis is testable. A test would consist
of seeing whether institutionally respected biologists could find
biomolecular information in the hallucinatory world of ayahuasqueros. However, this hypothesis is currently not receivable by
institutional biology, because it impinges on the discipline's presuppositions.

Biology has a blind spot of historical origin.

MY HYPOTHESIS SUGGESTS that what scientists call DNA corresponds to the animate essences that shamans say communicate
with them and animate all life forms. Modern biology, however, is
founded on the notion that nature is not animated by an intelligence and therefore cannot communicate.

This presupposition comes from the materialist tradition established by the naturalists of the eighteenth and nineteenth centuries. In those days, it took courage to question the explanations about life afforded by a literal reading of the Bible. By adopting a scientific method based on direct observation and the classification of species, Linnaeus, Lamarck, Darwin, and Wallace audaciously concluded that the different species had evolved over time—and had not been created in fixed form six thousand years previously in the Garden of Eden.

Wallace and Darwin simultaneously proposed a material mechanism to explain the evolution of species. According to their theory of natural selection, organisms presented slight variations from one generation to the next, which were either retained or eliminated in the struggle for survival. This idea rested on a circular argument: Those who survive are the most able to survive. But it seemed to explain both the variation of species and the astonishing perfection of the natural world, as it retained only the improvements. Above all, it took God out of the picture and enabled biologists to study nature without having to worry about a divine plan within.

For almost a century, the theory of natural selection was contested. Vitalists, like Bergson, rejected its stubborn materialism, pointing out that it lacked a mechanism to explain the origin of the variations. It wasn't until the 1950s and the discovery of the role of DNA that the theory of natural selection became generally accepted among scientists. The DNA molecule seemed to demonstrate the materiality of heredity and to provide the missing mechanism. As DNA is self-duplicating and transmits its information to proteins, biologists concluded that information could not flow back from proteins to DNA; therefore, genetic variation could only come from *errors* in the duplication process. Francis

Crick termed this the "central dogma" of the young discipline called molecular biology. "Chance is the only true source of novelty," he wrote.[1]

The discovery of DNA's role and the formulation in molecular terms of the theory of natural selection gave a new impetus to materialist philosophy. It became possible to contend on a scientific basis that life was a purely material phenomenon. Francis Crick wrote: "The ultimate aim of the modern movement in biology is to explain *all* biology in terms of physics and chemistry" (original italics). François Jacob, another Nobel Prize–winning molecular biologist, wrote: "The processes which occur in living beings at the microscopic level of molecules are in no way different from those analyzed by physics and chemistry in inert systems."[2]

The materialist approach in molecular biology went from strength to strength—but it rested on the unprovable presupposition that chance is the only source of novelty in nature, and that nature is devoid of any goal, intention, or consciousness. Jacques Monod, also a Nobel Prize–winning molecular biologist, expressed this idea clearly in his famous essay *Chance and necessity*: "The cornerstone of the scientific method is the postulate that nature is objective. In other words, the *systematic* denial that 'true' knowledge can be reached by interpreting phenomena in terms of final causes—that is to say, of 'purpose'. . . . This pure postulate is impossible to demonstrate, for it is obviously impossible to imagine an experiment proving the *nonexistence* anywhere in nature of a purpose, or a pursued end. But the postulate of objectivity is consubstantial with science, and has guided the whole of its prodigious development for three centuries. It is impossible to escape it, even provisionally or in a limited area, without departing from the domain of science itself"[3] (original italics).

Biologists thought they had found the truth, and they did not hesitate to call it "dogma." Strangely, their newfound conviction was hardly troubled by the discovery in the 1960s of a genetic code that is the same for all living beings and that bears striking similarities to human coding systems, or languages. To transmit information, the genetic code uses elements (A, G, C, and T) that are meaningless individually, but that form units of significance when combined, in the same way that letters make up words. The genetic code contains 64 three-letter "words," all of which have meaning, including two punctuation marks.

As linguist Roman Jakobson pointed out, such coding systems were considered up until the discovery of the genetic code as "exclusively human phenomena"[4]—that is, phenomena that require the presence of an intelligence to exist.

WHEN I STARTED READING the literature of molecular biology, I was stunned by certain descriptions. Admittedly, I was on the lookout for anything unusual, as my investigation had led me to consider that DNA and its cellular machinery truly were an extremely sophisticated technology of cosmic origin. But as I pored over thousands of pages of biological texts, I discovered a world of science fiction that seemed to confirm my hypothesis. Proteins and enzymes were described as "miniature robots," ribosomes were "molecular computers," cells were "factories," DNA itself was a "text," a "program," a "language," or "data." One only had to do a literal reading of contemporary biology to reach shattering conclusions; yet most authors display a total lack of astonishment and seem to consider that life is merely "a normal physicochemical phenomenon."[5]

One of the facts that troubled me most was the astronomical

length of the DNA contained in a human body: 125 billion miles. There, I thought, is the Ashaninca's sky-rope: It is inside us and is certainly long enough to connect earth and heaven. What did biologists make of this cosmic number? Most of them did not even mention it, and those who did talked of a "useless but amusing fact."

I was also troubled by the certitude exhibited by most biologists in the face of the profoundly mysterious reality they were describing. After all, the spectacular accomplishments of molecular biology during the second half of the twentieth century had led to more questions than answers. This is an old problem: Knowledge calls for more knowledge, or, as Jean Piaget wrote, "The most developed science remains a continual becoming."[6] Yet few biological texts discuss the unknown.

Take proteins, for instance. These long chains of amino acids, strung together in the order specified by DNA, accomplish almost all the essential tasks in cells. They catch molecules and build them into cellular structures or take them apart to extract their energy. They carry atoms to precise places inside or outside the cell. They act as pumps or motors. They form receptors that trap highly specific molecules or antennae that conduct electrical charges. Like versatile marionettes, or jacks-of-all-trades, they twist, fold, and stretch into the shape their task requires. What is known, precisely, about these "self-assembling machines"? According to Alwyn Scott, a mathematician with an interest in molecular biology: "Biologists' understanding of how proteins function is a lot like your and my understanding of how a car works. We know you put in gas, and the gas is burned to make things turn, but the details are all pretty vague."[7]

Enzymes are large proteins that accelerate cellular activities. They act with disarming speed and selectivity. One enzyme in human blood, carbonic anhydrase, can assemble single-handedly over a half million molecules of carbonic acid per second. The enzymes which both repair the double helix in case of damage and correct any errors in the DNA replication process make only one mistake every ten billion letters. Enzymes read the DNA text, transcribe it into RNA, edit out the non-coding passages, splice together the final message, construct the machines that read the instructions and build . . . other enzymes. What is known, precisely, about these "molecular automata"? According to biologists Chris Calladine and Horace Drew: "These enzymes are extremely efficient in doing their job, yet no one knows exactly how they work."[8]

Shamans say the correct way to talk about spirits is in metaphors. Biologists confirm this notion by using a precise array of anthropocentric and technological metaphors to describe DNA, proteins, and enzymes. DNA is a *text*, or a *program*, or *data*, containing *information*, which is *read* and *transcribed* into *messenger*-RNAs. The latter feed into ribosomes, which are *molecular computers* that *translate* the *instructions* according to the genetic *code*. They *build* the rest of the cell's *machinery*, namely the proteins and enzymes, which are *miniaturized robots* that construct and maintain the cell.

Over the course of my readings, I constantly wondered how nature could be devoid of intention if it truly corresponded to the descriptions biologists made of it.

One only had to consider the "dance of the chromosomes" to see DNA *move* in a deliberate way. During cell division, chromosomes double themselves and assemble by pairs. The two sets of chromosomes then line up along the middle of the cell and migrate toward their respective pole, each member of each pair always going in the direction opposite to its companion's. How could this "amazing, stately pavane"[9] occur without some form of intention?

In biology, this question is simply not asked. DNA is "just a chemical,"[10] deoxyribonucleic acid, to be precise. Biologists describe it as both a molecule and a language, making it the informational substance of life, but they do not consider it to be conscious, or alive, because chemicals are inert by definition.

How, I wondered, could biology presuppose that DNA is not conscious, if it does not even understand the human brain, which is the seat of our own consciousness and which is built according to the instructions in our DNA? How could nature not be conscious if our own consciousness is produced by nature?[11]

As I patrolled the texts of biology, I discovered that the natural world was teeming with examples of behaviors that seem to require *forethought*. Some crows manufacture tools with standardized hooks and toothed probes to help in their search for insects hidden in holes. Some chimpanzees, when infected with intestinal parasites, eat bitter, foul-tasting plants, which they otherwise avoid and which contain biologically active compounds that kill intestinal parasites. Some species of ants, with brains the size of a grain of sugar, raise herds of aphids which they milk for their sweet secretions and which they keep in barns. Other ants have been cultivating mushrooms as their exclusive food for fifty million years.[12] It is difficult to understand how these insects could do this without a form of consciousness. Yet scientific observers deny them this faculty, like Jacques Monod, who considers the behavior of bees to be "automatic": "We know the hive is 'artificial' in so far as it represents the product of the activity of the bees. But we have good reasons for thinking that this activity is strictly automatic—immediate, but not consciously planned."[13]

Indeed, the "postulate of objectivity" prevents its practitioners from recognizing any intentionality in nature or, rather, it nullifies their claim to science if they do so.

DURING THIS INVESTIGATION, I became familiar with certain limits of the rational gaze: It tends to fragment reality and to exclude complementarity and the association of contraries from its field of vision. I also discovered one of its more pernicious effects: The rational approach tends to minimize what it does not understand.

Anthropology is an ideal training ground for learning this. The first anthropologists went out beyond the limits of the rational world and saw *primitives* and *inferior* societies. When they met shamans, they thought they were *mentally ill*.

The rational approach starts from the idea that everything is explainable and that mystery is in some sense the enemy. This means that it prefers pejorative, and even wrong, answers to admitting its own lack of understanding.

The molecular biology that considers that 97 percent of the DNA in our body is "junk" reveals not only its degree of ignorance, but the extent to which it is prepared to belittle the unknown. Some recent hypotheses suggest that "junk DNA" might have certain functions after all.[14] But this does not hide the pejorative reflex: We don't understand, so we shoot first, then ask questions. This is cowboy science, and it is not as objective as it claims. Neutrality, or simple honesty, would have consisted in saying "for the moment, we do not know." It would have been just as easy to call it *mystery* DNA, for instance.

The problem is not having presuppositions, but failing to make them explicit. If biology said about the intentionality that nature seems to manifest at all levels, "we see it sometimes, but cannot discuss it without ceasing to do science according to our own criteria," things would at least be clear. But biology tends to project its presuppositions onto the reality it observes, claiming that nature itself is devoid of intention.

This is perhaps one of the most important things I learned during this investigation: We see what we believe, and not just the contrary; and to change what we see, it is sometimes necessary to change what we believe.

AT FIRST I THOUGHT I was the only one to realize that biology had limits similar to those of scientific anthropology and that it, too, was a "self-flattering imposture," which treats the living as if it were inert. Then I discovered that there were all sorts of people within the scientific community who were already discussing biology's fundamental contradictions.

During the 1980s, it became possible to determine the exact sequence of amino acids in given proteins. This revealed a new level of complexity in living beings. A single nicotinic receptor, forming a highly specific lock coupled to an equally selective channel, is made of five juxtaposed protein chains that contain a total of 2,500 amino acids lined up in the right order. Despite the improbability of the chance emergence of such a structure, even nematodes, which are among the most simple multicellular invertebrates, have nicotinic receptors.[15]

Confronted by this kind of complexity, some researchers no longer content themselves with the usual explanation. Robert Wesson writes in his book *Beyond natural selection:* "No simple theory can cope with the enormous complexity revealed by modern genetics."[16]

Other researchers have pointed out the improbability of the mechanism that is supposed to be the source of variation— namely, the accumulation of errors in the genetic text. It seems obvious that "a message would quickly lose all meaning if its contents changed continuously in an anarchic fashion."[17] How, then, could such a process lead to the prodigies of the natural world, of which we are a part?

Another fundamental problem contradicts the theory of chance-driven natural selection. According to the theory, species should evolve slowly and gradually, since evolution is caused by the accumulation and selection of random errors in the genetic text. However, the fossil record reveals a completely different scenario. J. Madeleine Nash writes in her review of recent research in paleontology: "Until about 600 million years ago, there were no organisms more complex than bacteria, multicelled algae and single-celled plankton. . . . Then, 543 million years ago, in the early Cambrian, within the span of no more than 10 million years, creatures with teeth and tentacles and claws and jaws materialized with the suddenness of apparitions. In a burst of creativity like nothing before or since, nature appears to have sketched out the blueprints for virtually the whole of the animal kingdom. . . . Since 1987, discoveries of major fossil beds in Greenland, in China, in Siberia, and now in Namibia have shown that the period of biological innovation occurred at virtually the same instant in geological time all around the world. . . . Now, . . . virtually everyone agrees that the Cambrian started almost exactly 543 million years ago and, even more startling, that all but one of the phyla in the fossil record appeared within the first 5 to 10 million years."[18]

Throughout the fossil record, species seem to appear suddenly, fully formed and equipped with all sorts of specialized organs, then remain stable for millions of years. For instance, there is no intermediate form between the terrestrial ancestor of the whale and the first fossils of this marine mammal. Like their current descendants, the latter have nostrils situated atop their heads, a modified respiratory system, new organs like a dorsal fin, and nipples surrounded by a cap to keep out seawater and equipped with a pump for underwater suckling.[19] The whale represents the rule, rather than the exception. According to biologist Ernst Mayr, an authority on the matter of evolution, there is "no clear evidence for any change of a species into a different genus or for the gradual origin of an evolutionary novelty."[20]

A similar problem exists at the cellular level. Microbiologist James Shapiro writes: "In fact, there are no detailed Darwinian accounts for the evolution of any fundamental biochemical or cellular system, only a variety of wishful speculations. It is remarkable that Darwinism is accepted as a satisfactory explanation for such a vast subject—evolution—with so little rigorous examination of how well its basic theses work in illuminating specific instances of biological adaptation or diversity."[21]

In the middle of the 1990s, biologists sequenced the first complete genomes of free-living organisms. So far, the smallest known bacterial genome contains 580,000 DNA letters.[22] This is an enormous amount of information, comparable to the contents of a small telephone directory. When one considers that bacteria are the smallest units of life as we know it, it becomes even more difficult to understand how the first bacterium could have taken form spontaneously in a lifeless, chemical soup. How can a small telephone directory of information emerge from random processes?

The genomes of more complex organisms are even more daunting in size. Baker's yeast is a unicellular organism that contains 12 million DNA letters; the genome of nematodes, which are rather simple multicellular organisms, contains 100 million DNA letters. Mouse genomes, like human genomes, contain approximately 3 billion DNA letters.

By mapping, sequencing, and comparing different genomes, biologists have recently found further levels of complexity. Some sequences are highly conserved between species. For example, 400 human genes match very similar genes in yeast. This means these genes have stayed in a nearly identical place and form over hundreds of millions of years of evolution, from a very primitive form of life to a human being.[23]

Some genetic sequences, known as "master genes," control hundreds of other genes like an on/off switch. These master genes also seem to be highly conserved across species. For example, flies and human beings have a very similar gene that controls the development of the eye, though their eyes are very different. Geneticist André Langaney writes that the existence of master genes "points to the insufficiency of the neodarwinian model and to the necessity of introducing into the theory of evolution mechanisms, either known or to be discovered, that contradict this model's basic principles."[24]

Recent gene mapping has revealed that, in some areas of the DNA text, genes are thirty times more dense than in other areas, and some of the genes appear to clump together in families that work on similar problems. In some cases, gene clumps are highly conserved across species, as in the X chromosome of mice and humans, for example. In both species, the X chromosome is a giant molecule of DNA, some 160 million nucleotides long; it is one of the pair of chromosomes that determine whether an offspring is male or female. The mapping of the X chromosome has shown that genes are bunched together mostly in five gene-rich regions, with lengthy, apparently desert regions of DNA in between, and that mice and humans have much the same set of genes on their X chromosomes even though the two species have followed separate evolutionary paths for 80 million years.[25]

Recent work on genetic sequences is starting to reveal much greater complexity than could have been conceived even ten years previous to the data's emergence. How are scientists going to make sense of the overwhelming complexity of DNA texts? Robert Pollack proposes "that DNA is not merely an informational molecule, but is also a form of text, and that therefore it is best understood by analytical ways of thinking commonly applied to other forms of text, for example, books."[26] This seems to be a sensible suggestion, but it begs the question: How can one analyze a text if one presupposes that no intelligence wrote it?

Despite these essential contradictions, which I sum up here in a few lines but which could fill entire books, the theory of natural selection remains firmly in place in the minds of most biologists. This is because it is always possible to claim that the appropriate mutations occurred by chance and were selected. But this undemonstrable proposition is denounced by an increasing number of scientists. Pier Luigi Luisi talks of the "tautology of molecular Darwinism . . . [which] is unable to elicit concepts other than those from which it has been originally constructed."

The circularity of the Darwinian theory means that it is not falsifiable and therefore not truly scientific. The "falsifiability criterion" is the cornerstone of twentieth-century scientific method. It was developed by philosopher Karl Popper, who argued that one could never prove a scientific theory to be correct, because only an infinite number of confirming results would constitute definitive proof. Popper proposed instead to test theories in ways that seek to contradict, or falsify, them; the absence of contradictory evidence thereby becomes proof of the theory's validity. Popper writes: "I have come to the conclusion that Darwinism is not a testable scientific theory, but a *metaphysical research pro-*

gramme—a possible framework for testable scientific theories. . . . It is metaphysical because it is not testable"[27] (original italics).

Biology is currently divided between a majority who consider the theory of natural selection to be true and established as fact and a minority who question it.

However, the critics of natural selection have yet to come up with a new theory to replace the old one and institutions sustain current orthodoxies by their inertia. A new biological paradigm is still a long way off.

PRESUPPOSITIONS, POSTULATES, and circular arguments pertain more to faith than to science. My approach in this book starts from the idea that it is of utmost importance to respect the faith of others, no matter how strange, whether it is shamans who believe plants communicate or biologists who believe nature is inanimate.

I do not intend to attack anybody's faith, but to demarcate the blind spot of the rational and fragmented gaze of contemporary biology and to explain why my hypothesis is condemned in advance to remain in that spot. To sum up: My hypothesis is based on the idea that DNA in particular and nature in general are minded. This contravenes the founding principle of the molecular biology that is the current orthodoxy.

"WHAT TOOK YOU SO LONG?"

In Rio, the governments of the world signed treaties that recognize the ecological knowledge of indigenous people, as well as the importance of compensating it "equitably." However, as I think I have shown in this book, the scientific community is not ready to engage in a true dialogue with indigenous people, as biology cannot receive their knowledge owing to several epistemological blocks.

Paradoxically, this is an advantage for indigenous people, because it gives them time to prepare themselves. If the hypothesis presented in this book is correct, it means that they have not only a precious understanding of specific plants and remedies, but *an unsuspected source of biomolecular knowledge,* which is financially invaluable and mainly concerns tomorrow's science.

I will continue working with the indigenous organizations of the Amazon and will discuss with them the possible consequences of my hypothesis. I will tell them that biology has become an industry that is guided by a thirst for marketable knowledge, rather than by ethical and spiritual considerations.

It will be up to them to decide which strategy to adopt.

Perhaps they will simply try to cash in on their knowledge, by learning about molecular biology and then looking for marketable biological information in the shamanic sphere. After all, the fact that current biology cannot receive indigenous knowledge has not stopped pharmaceutical companies from commercializing parts of it.

Over the last five hundred years, the Western world has demonstrated that it is in no hurry to compensate the knowledge of indigenous people, even though it has used this knowledge repeatedly. The years that have gone past since the Rio treaties have changed nothing in this regard. Under these circumstances, I can only think of advising indigenous organizations to negotiate a hard line.

To start with, this would mean increasing controls on the scientists who wish to gain access to their shamanism. In a world governed by money and the race to success, where everything is patentable and marketable (including DNA sequences), it is important to play the game like everybody else and to protect one's trade secrets.

However, it does not seem probable that molecular biologists will be able to steal the secrets of ayahuasqueros in the near future. There is more to becoming a Western Amazonian shaman than just drinking ayahuasca. One must follow a long and terrifying apprenticeship based on the repeated ingestion of hallucinogens, prolonged diets, and isolation in the forest to master one's hallucinations. This does not seem to be within the reach of most Westerners.[1] I, for one, would be incapable of it.

Furthermore, Western culture does not facilitate such an apprenticeship; it considers the main hallucinogenic plants illegal, and most "recreational" users, who disregard the law, fail to practice the required techniques (fasting, abstinence from alcohol and sex, darkness, chanting, etc.). To my mind, a truly hallucinatory session is more like a controlled nightmare than a form of recreation and demands know-how, discipline, and courage.

THROUGHOUT THIS BOOK, my approach has consisted of translating the shamanism of ayahuasqueros to make it understandable to a Western audience. I believe it is in the interest of Amazonia's indigenous people that their knowledge be understood in Western terms, because the world is currently governed by Western values and institutions. For instance, it was not until Western countries realized that it was in their own interest to protect tropical forests that it became possible to find the funds to demarcate the territories of the indigenous people living there. Prior to that, most territorial claims, formulated in terms of the indigenous people's own interests, led to nothing.

My conclusion can be accused of reductionism, as I end up presenting in mainly biological terms practices that simultaneously combine music, cosmology, hallucinations, medicine, botany, and psychology, among others.[2] My interpretation, focusing on molecular biology, certainly distorts shamanism's multidimensionality, but it at least attempts to bring together a number of compartmentalized disciplines, from mythology to neurology through anthropology and botany. I do not mean that shamanism is equivalent to molecular biology, but that for us fragmented Westerners, molecular biology is the most fruitful approach to the holistic reality of shamanism, which has become so unfamiliar to us.

ELEVEN YEARS AGO, I arrived for the first time as a young anthropologist in the Ashaninca village of Quirishari and quickly struck a deal with its inhabitants. They would allow me to live with them and to study their practices so that I could explain them to the people in my country and become a doctor of anthropology. In exchange I was to teach them an "accounting" course—that is, arithmetic. Their position was clear: An anthropologist should not only study people, but try to be *useful* to them as well.

Carlos Perez Shuma, who took me under his wing, often explained my presence to his companions by saying, "He has come to live with us for two years because he wants to tell the people in his country how we work." These people had always been told by missionaries, colonists, and governmental agronomists that they knew nothing—and that their so-called ignorance even justified the confiscation of their lands. So they were not displeased at the idea of demonstrating their knowledge. This is the license on the basis of which I wrote this book.

All the Ashaninca I met wanted to participate in the world market, if only to acquire the commodities that make life easier in the rainforest, such as machetes, axes, knives, cooking pots, flashlights, batteries, and kerosene. They also needed money to meet the minimal requirements of "civilization," namely clothes, schoolbooks, pens and paper, and everyone dreamed of owning a radio or a tape recorder.

Beyond money and commodities, the indigenous people of the Amazon aspire to survival in a world that has considered them, until recently, as little more than Stone Age savages. Now they all demand the demarcation and titling of their territories, as well as the means to educate their children in their own terms.

Western institutions seem finally to have understood, at least in principle, the importance of recognizing indigenous territories—though much remains to be done in practice on this count.

However, the indigenous claim to bilingual and intercultural education has yet to be heard, even though it would seem to be a prerequisite to the establishment of a truly rational dialogue with these people. After all, the word "rational" comes from the Latin *ratio*, "calculation." How can one establish the "equitable" compensation of indigenous knowledge if the majority of indigenous people do not understand the basics of accounting and money management and require training in arithmetic?

This is not a gratuitous question. Research has shown that Western-style education does not work with Amazonian Indians. Theirs is an oral tradition, where knowledge is mainly acquired through practice in nature. When one shuts young Indians into a schoolroom for six hours a day, nine months a year, for ten years, and teaches them foreign concepts in a European language, they end up reaching, on average, a level of second-grade primary school. This means that most of them barely know how to read and write and do not know how to calculate a percentage.

The indigenous people themselves are the first to realize what a disadvantage this gives them in a world defined by written words and numbers. Practically speaking, they know that they are often shortchanged when they sell their products on the market. This is why they want bilingual and intercultural education. However, for each indigenous society, speaking its own language, it is necessary to develop a specific curriculum and to train indigenous instructors capable of teaching it.[3] This costs approximately U.S.$200,000 per culture. In the Peruvian Amazon alone, there are fifty-six different cultures, each speaking a different language. For the moment, only ten of these have access to bilingual, intercultural education. Why so few? Because the small number of nongovernmental organizations supporting this initiative have limited means, and the institutions that are large enough to fund education programs for indigenous people seem to be in no hurry to do so. It is true that the results of such an investment can only be measured in generations, rather than in five-year periods.

AFTER WRITING the original French version of this book, I returned to the Peruvian Amazon and spent a week in Iquitos at a school for bilingual, intercultural education, where young men and women from ten indigenous societies are learning to teach both indigenous and Western knowledge both in their mother tongue and in Spanish. I spent several fascinating days observing from the back of a class, then the students asked whether I would tell them about my work. On my last evening I addressed a roomful of students and told them my hypothesis indicated there was a relationship between the entwined serpents Amazonian shamans see in their visions and the DNA double helix that science discovered in 1953. At the end of the talk, a voice called out from the back: "Are you saying that scientists are catching up with us?"

I also returned to Quirishari and met with Carlos Perez Shuma for the first time in nine years. He hadn't changed at all, and even seemed younger. We sat down in a quiet house and began to chat, making up for lost time. He told me about all the things that had occurred in the Pichis Valley during my absence. I listened to him for about an hour, but then could no longer contain myself: "Uncle," I said, "there is something important I have to tell you. You remember all those things you explained into the tape recorder that I had difficulty understanding? Well, after thinking about it for years, and then studying it, I have just discovered that in scientific terms all the things you told me were true." I thought he would be pleased and was about to continue when he interrupted. "What took you so long?" he said.

WE WESTERNERS have our paradoxes. Rationalism has brought us unhoped-for material well-being, yet few people seem satisfied.

However, we are not alone, and indigenous people also have their dilemmas.

First, in order to recognize the true value of their knowledge, they must face the loss history has inflicted on them. For the last 500 years, Western civilization has been teaching indigenous people that they know nothing—to the point that some of them have come to believe it. For them to appreciate the value of their own knowledge, they must come to terms with having been misled.

Second, there is money. Over the last few years, one of the main problems confronting the indigenous organizations of Amazonia has been their own success. Friends of the rainforest have poured money into the area, with the best intentions, but without rigorous controls. This has mainly caused corruption and division. The fault is also ours, because we trusted them in a paternalistic fashion. We thought that indigenous people were incorruptible, because we had romantic presuppositions. But this does not mean we should stop working with them; rather, we should insist on greater controls in the management of funds to avoid counterproductive generosity that draws its roots in romantic paternalism.

Finally, the creation of compensation mechanisms for the intellectual property of indigenous people will depend on the resolution of the following dilemma. In shamanic traditions, it is invariably specified that spiritual knowledge is not marketable. Certainly, the shaman's work deserves remuneration, but, by definition, the sacred is not for sale; the use of this knowledge for the accumulation of personal power is the definition of black magic.

In a world where everything is for sale, including genetic sequences, this concept will no doubt be difficult to negotiate.[4]

I SPEAK OF "indigenous people," or "Amazonian Indians," and I oppose them to "us Westerners"—yet these words do not correspond to monolithic realities. Prior to European colonization, the inhabitants of the Amazon already made up a patchwork of diversity, with hundreds of cultures speaking different languages and enjoying more or less constructive relations among each other. Some indigenous societies did not wait for the conquistadores' arrival to wage war on each other.

The diversified reality of indigenous Amazonia was assaulted by European colonization, which decimated the population and fragmented territories. Indigenous cultures survive, strong here, less so there, necessarily transformed and hybridized. But appearances are misleading, and reality is often double-edged: Hybridization, mestizo-ization, which implies a certain dilution, is one of the oldest survival strategies in the world. The "true Indian" who has never left the forest, does not speak a word of Spanish or Portuguese, uses no metal tools, and wanders around naked and feathered exists only in the Western imagination. Which is just as well for the real-life Indians, because they already have a hard enough time leading their lives as they see fit.

Ayahuasca-based shamanism is essentially an indigenous phenomenon. However, it is also true that this shamanism is currently enjoying a boom thanks to the mixing of cultures. The case of Pablo Amaringo is eloquent in this respect. Amaringo is a mestizo ayahuasquero. He lives in the town of Pucallpa, his mother tongue is Quechua, and his ancestry is a mix of Cocama, Lamista, and Piro. The songs he sings in his hallucinatory trances have indigenous lyrics. Amaringo does not consider himself an Indian, though he recognizes the indigenous nature of his knowledge. For instance, he says the Ashaninca are the ones who know "better than any other jungle people the magical uses of plant-teachers."[5]

Meanwhile, the Ashaninca people I knew in the Pichis claimed that the best shamans were Shipibo-Conibo (who live in the same area as Amaringo). Ruperto Gomez, the ayahuasquero who initiated me, did his apprenticeship with the Shipibo-Conibo, and this conferred undeniable prestige on him. So it would seem that studies "abroad" are considered better and that the high place of Amazonian shamanism is always somewhere other than where one happens to be.[6]

Shamanism resembles an academic discipline (such as anthropology or molecular biology); with its practitioners, fundamental researchers, specialists, and schools of thought it is a way of apprehending the world that evolves constantly. One thing is certain: Both indigenous and mestizo shamans consider people like the Shipibo-Conibo, the Tukano, the Kamsá, and the Huitoto as the equivalents to universities such as Oxford, Cambridge, Harvard, and the Sorbonne[7]; they are the highest reference in matters of knowledge. In this sense, ayahuasca-based shamanism is an essentially indigenous phenomenon. It belongs to the indigenous people of Western Amazonia, who hold the keys to a way of knowing that they have practiced without interruption for at least five thousand years. In comparison, the universities of the Western world are less than nine hundred years old.

The shamanism of which the indigenous people of the Amazon are the guardians represents knowledge accumulated over thousands of years in the most biologically diverse place on earth. Certainly, shamans say they acquire their knowledge directly from the spirits, but they grow up in cultures where shamanic visions are stored in myths. In this way mythology informs shamanism: The invisible, life-creating maninkari spirits are the ones whose feats Ashaninca mythology relates, and it is also the maninkari who talk to Ashaninca shamans in their visions and tell them how to heal.

An indigenous culture with sufficient territory, and bilingual and intercultural education, is in a better position to maintain and cultivate its mythology and shamanism. Conversely, the confiscation of their lands and imposition of foreign education, which turns their young people into amnesiacs, threatens the survival not only of these people, but of an entire way of knowing. It is as if one were burning down the oldest universities in the world and their libraries, one after another—thereby sacrificing the knowledge of the world's future generations.

IN THIS BOOK I chose an autobiographical and narrative approach for several reasons. First, I do not believe in an objective point of view with an exclusive monopoly on reality. So it seemed important to expose the inevitable presuppositions that any observer has, so that readers may come to their opinion in full knowledge of the setting.[8]

In this sense I belong to the recent movement within anthropology that views the discipline as a form of interpretation rather than as a science. However, even among my colleagues who work in this fashion, listening to people carefully, recording and transcribing their words, and interpreting them as well as they can, there remains a problem I have tried to avoid—namely, the compartmentalization of knowledge into disciplines, which means that the discourse of a given specialist is only understandable to his or her immediate colleagues.[9] In my opinion, subjects such as DNA and the knowledge of indigenous people are too important to be entrusted solely to the focalized gaze of academic specialists in biology or anthropology; they concern indigenous people themselves, but also midwives, farmers, musicians, and all the rest. I decided to tell my story in an attempt to create an account that would be comprehensible across disciplines and outside the academy.

This decision was inspired by shamanic traditions, which invariably state that images, metaphors, and stories are the best means to transmit knowledge. In this sense, myths are "scientific narratives," or stories about knowledge (the word "science" comes from the Latin *scire,* "to know").

I was fortunate to choose this approach, because it was in telling my story that I discovered the real story I wanted to tell.

There was a price to pay for implicating myself in my work like this. I spent many sleepless nights and put a strain on my personal life. I was truly bowled over by working on this book. At the time, I felt sure it was going to change the world. It took months of talking with numerous friends to understand that my hypothesis was not even receivable by official science, despite the scientific elements it contains. Since then, I've calmed down and no longer talk away for hours.

We live in a time when it is difficult to speak seriously about one's spirituality. Often one only has to state one's convictions to be considered a preacher. I, too, support the idea that everybody should be free to believe what they want and that it is nobody's business to tell others what they should believe. So I will not describe in detail the impact of my work on my own spirituality, and I will not tell readers what to think about the connections I have established.

Here, too, I draw my inspiration from shamanism, which rests not on doctrine, but on *experience*. The shaman is simply a guide, who conducts the initiate to the spirits. The initiate picks up the information revealed by the spirits and does what he or she wants with it. Likewise, in this book, I provide a number of connections, with complete references for those who wish to follow a particular trail. In the end, it is up to the readers to draw the spiritual conclusions they see fit.

Is there a goal to life? Do we exist for a reason? I believe so, and I think that the combination of shamanism and biology gives interesting answers to these questions. But I do not feel ready to discuss them from a personal point of view.

The microscopic world of DNA, and its proteins and enzymes, is teeming inside us and is enough to make us marvel. Yet rational discourse, which holds a monopoly on the subject, denies itself a sense of wonder. Current biologists condemn themselves, through their beliefs, to describe DNA and the cell-based life for which it codes as if they were blind people discussing movies or objective anthropologists explaining the hallucinatory sphere of which they have no experience: They oblige themselves to consider an animate reality as if it were inanimate.

By ignoring this obligation, and by considering shamanism and biology at the same time, stereoscopically, I saw DNA snakes. They were alive.

THE ORIGIN OF KNOWLEDGE is a subject that anthropologists neglect—which is one of the reasons that prompted me to write this book. However, anthropologists are not alone; scientists in general seem to have a similar difficulty. On closer examination, the reason for this becomes obvious: Many of science's central ideas seem to come from beyond the limits of rationalism. René

Descartes dreams of an angel who explains the basic principles of materialist rationalism to him; Albert Einstein daydreams in a tram, approaching another, and conceives the theory of relativity; James Watson scribbles on a newspaper in a train, then rides his bicycle to reach the conviction (having "borrowed" Rosalind Franklin's radiophotographic work) that DNA has the form of a double helix.[10] And so on.

Scientific discovery often originates from a combination of focalized and defocalized consciousness. Typically, a researcher spends months in the lab working on a problem, considering the data to the point of saturation, then attains illumination while jogging, daydreaming, lying in bed making mental pictures, driving a car, cooking, shaving, bathing—in brief, while thinking about something else and *defocalizing*. W. I. B. Beveridge writes in *The art of scientific investigation:* "The most characteristic circumstances of an intuition are a period of intense work on the problem accompanied by a desire for its solution, abandonment of the work perhaps with attention to something else, then the appearance of the idea with dramatic suddenness and often a sense of certainty. Often there is a feeling of exhilaration and perhaps surprise that the idea had not been thought of previously."[11]

During this investigation I complemented months of straightforward scholastic work (reading, note taking, and categorizing) with defocalized approaches (such as walking in nature, nocturnal soliloquies, dissonant music, daydreaming), which greatly helped me find my way. My inspiration for this is once again shamanic. But shamans are not the only ones to seek knowledge by cultivating defocalization. Artists have done this throughout the ages. As Antonin Artaud wrote: "I abandon myself to the fever of dreams, in search for new laws."[12]

DID I SEE imaginary connections in my fever? Am I wrong in linking DNA to these cosmic serpents from around the world, these sky-ropes and axis mundi? Some of my colleagues will think so. Here's one of the reasons:

In the nineteenth century the first anthropologists set about comparing cultures and elaborating theories on the basis of the similarities they found. When they discovered, for instance, that bagpipes were played not only in Scotland, but in Arabia and the Ukraine, they established false connections between these cultures. Then they realized that people could do similar things for different reasons. Since then, anthropology has backed away from grand generalizations, denounced "abuses of the comparative method," and locked itself into specificity bordering on myopia. This is why anthropologists who study Western Amazonia's hallucinatory shamanism limit themselves to specific analyses of a given culture—failing to see the essential common points between cultures. So their fine-grained analyses allow them to see that the diet of an apprentice ayahuasquero is based on the consumption of bananas and/or fish. But they do not notice that this diet is practiced throughout Western Amazonia, and so they do not consider that it may have a biochemical basis—which in fact it does.

By shunning comparisons between cultures, one ends up masking true connections and fragmenting reality a little more, without even realizing it.

Is the cosmic serpent of the Shipibo-Conibo, the Aztecs, the Australian Aborigines, and the Ancient Egyptians the same? No, will reply the anthropologists who insist on cultural specificity; to believe otherwise, according to them, comes down to making the same mistake as Mircea Eliade four decades ago, when he detached all those symbols from their contexts, obliterated the sociocultural aspect of phenomena, mutilated the facts, and so on. The critique is well known now, and it is time to turn it on its head. In the name of *what* does one mask fundamental similarities in human symbolism—if not out of a stubborn loyalty to rationalist fragmentation? How can one explain these similarities with a concept other than chance—which is more an *absence of concept* than anything? Why insist on taking reality apart, but never try putting it back together again?

ACCORDING TO MY HYPOTHESIS, shamans take their consciousness down to the molecular level and gain access to biomolecular information. But what actually goes on in the brain/mind of an ayahuasquero when this occurs? What is the nature of a shaman's communication with the animate essences of nature? The clear answer is that more research is needed in consciousness, shamanism, molecular biology, and their interrelatedness.

Rationalism separates things to understand them. But its fragmented disciplines have limited perspectives and blind spots. And as any driver knows, it is important to pay attention to blind spots, because they can contain vital information. To reach a fuller understanding of reality, science will have to shift its gaze. Could shamanism help science to defocalize? My experience indicates that engaging shamanic knowledge requires looking into a great number of disciplines and thinking about how they fit together.

FINALLY, a last question: Where does life come from?

Over the last decade, scientific research has come up against the impossibility that a single bacterium, representing the smallest unit of independent life as we know it, could have emerged by chance from any kind of "prebiotic soup."[13] Given that a cosmic origin, such as the one proposed by Francis Crick in his "directed panspermia" speculation, is not scientifically verifiable, scientists have focused almost exclusively on terrestrial scenarios.[14] According to these, precursor molecules took shape (by chance) and prepared the way for a world based on DNA and proteins. However, these different scenarios—based on RNA, peptides, clay, undersea volcanic sulfur, or small oily bubbles—all propose explanations relying on systems that have, by definition, been replaced by life as we know it, without leaving any traces.[15] These, too, are speculations that cannot be verified scientifically.[16]

The scientific study of the origins of life leads to an impasse, where agnosticism seems to be the only reasonable and rigorous position. As Robert Shapiro writes in his book *Origins: A skeptic's guide to the creation of life on Earth:* "We do not have the slightest idea about how life got started. The very particular set of chemicals that were necessary remains unknown to us. The process itself could have included an improbable event, as it could have happened according to a practically ineluctable sequence. It could have required several hundred million years, or only a few millennia. It could have happened in a tepid pool, or in a hydrothermal source at the bottom of the ocean, in a bubble in the atmosphere, or somewhere else than on Earth, out in the cosmos."[17]

Any certitude on this question is a matter of *faith*. So what do shamanic and mythological traditions say in this regard? According to Lawrence Sullivan, who has studied the indigenous religions of South America in detail: "In the myths recorded to date, the majority of South American cultures show little extended interest in absolute beginnings."[18]

Where does life come from? Perhaps the answer is not graspable by mere human beings. Chuang-Tzu implied as much a long time ago, when he wrote: "There is a beginning. There is a not yet beginning to be a beginning. There is a not yet beginning to be a not yet beginning to be a beginning. There is being. There is nonbeing. There is a not yet beginning to be nonbeing. There is a not yet beginning to be a not yet beginning to be nonbeing. Suddenly there is nonbeing. But I do not know, when it comes to nonbeing, which is really being and which is nonbeing. Now I have just said something. But I do not know whether what I have said has really said something or whether it hasn't said something."[19]

All things considered, wisdom requires not only the investigation of many things, but contemplation of the mystery.

NOTES

1: FOREST TELEVISION

1. According to La Barre (1976), an anthropologist known for his studies of the indigenous uses of the peyote cactus, Castaneda's first book "is pseudo-profound, sophomoric and deeply vulgar. To one reader at least, for decades interested in Amerindian hallucinogens, the book is frustratingly and tiresomely dull, posturing pseudo-ethnography and, intellectually, kitsch" (p. 42). De Mille (1980) calls Castaneda's work a "hoax" and a "farce" (pp. 11, 22).

2. The projects were carried out despite an independent evaluation done in 1981 for the United States Agency for International Development, which showed that all the "uninhabited" areas the Peruvian government proposed to develop and colonize were actually occupied by indigenous people who had been there for millennia and who, in some cases, had already reached their territory's carrying capacity—see Smith (1982, pp. 39–57).

3. A large majority of Ashaninca men living in the Pichis Valley in 1985 spoke fluent Spanish.

4. *Toé* is *Brugmansia suaveolens*. According to Schultes and Hofmann (1979, pp.128–129), *Brugmansia* and *Datura* were long considered to belong to the same genus, but were finally separated for morphological and biological reasons. However, their alkaloid content is similar.

2: ANTHROPOLOGISTS AND SHAMANS

1. In this paragraph, I simplify the possible ingredients of ayahuasca. Building on the work of Rivier and Lindgren (1972), McKenna, Towers, and Abbott (1984) show that the *Psychotria viridis* bush

(*chacruna* in Spanish) is almost invariably the source of the dimethyltryptamine contained in the ayahuasca brew prepared in the
Peruvian Amazon, while in Colombia the *Diplopterys cabrerana*
vine is used instead. The only constant in the different ayahuasca
recipes is the *Banisteriopsis caapi* vine, containing three
monoamine oxidase inhibitors, harmine, harmaline, and tetrahydro-
harmine, which are also hallucinogenic at sufficient dose levels. As
Luna (1986) points out, the basic mixture is often used to reveal the
properties of all sorts of other plants; thus, "the number of additives
is unlimited, simply because ayahuasca is a means of exploring properties of new plants and substances by studying the changes they
cause on the hallucinatory experience, and by examining the content of the visions" (p. 159). According to McKenna, Luna, and
Towers (1986), ayahuasca admixtures constitute a veritable "non-
investigated pharmacopoeia." It should also be noted that the *Banisteriopsis caapi* vine is commonly known as "ayahuasca," not to be
confused with the brew of the same name of which it is a component. See Schultes and Hofmann (1979) for further information on
these different plants. Concerning the endogenous production of
dimethyltryptamine in the human brain, see Smythies et al. (1979)
and Barker et al. (1981)—though Rivier (1996 personal communication) warns that current extraction procedures can lead to chemical transformation and that the presence of dimethyltryptamine in
extracted cerebrospinal liquid does not prove its endogenous existence; it could simply be the result of the transformation of endogenous tryptamines, such as 5-hydroxytryptamine (serotonin).
According to the archeological evidence gathered by Naranjo
(1986), Amazonian peoples have been using ayahuasca for at least
five thousand years. The quote in the text is from Schultes (1972, pp.
38–39). Finally, Lévi-Strauss (1950) writes: "Few primitive people
have acquired as complete a knowledge of the physical and chemical
properties of their botanical environment as the South American Indian" (p. 484).

2. The use of hallucinogens is by no means uniform across the immen-
sity of the Amazonian Basin. Out of approximately 400 indigenous
peoples, Luna (1986) lists 72 who use ayahuasca and who are concentrated in Western Amazonia. In other parts of the Amazon,
dimethyltryptamine-based hallucinogens are also used, but are extracted from different plants, such as *Virola*—which is snuffed in
powder form (see Schultes and Hofmann 1979, pp. 164–171). Some

peoples use only tobacco, the hallucinogenic properties of which have been documented by Wilbert (1987). Finally, in some Amazonian cultures, shamans work with dreams rather than hallucinations (see Perrin 1992b, Kracke 1992, and Wright 1992). See Schultes and Raffauf (1990, p. 9) for the estimate of 80,000 plant species in the Amazon.

3. Reichel-Dolmatoff (1971, 1975, 1978), Chaumeil (1982, 1983), Chevalier (1982), Luna (1984, 1986), and Gebhart-Sayer (1986) are exceptions.

4. Darwin (1871, p. 197).

5. The word "primitive" comes from the Latin *primitivus*, first born. Regarding the foundation of anthropology on an illusory object of study, see Kuper (1988).

6. Tylor (1866, p. 86). The word "savage" comes from the Latin *silvaticus*, "of the forest."

7. Malinowski (1922) writes with satisfaction: "Ethnology has introduced law and order into what seemed chaotic and freakish. It has transformed for us the sensational, wild and unaccountable world of 'savages' into a number of well ordered communities, governed by law, behaving and thinking according to consistent principles" (pp. 9–10).

8. Lévi-Strauss (1963a), explaining the notion of "order of orders," writes: "Thus anthropology considers the whole social fabric as a network of different types of orders. The kinship system provides a way to order individuals according to certain rules; social organization is another way of ordering individuals and groups; social stratifications, whether economic or political, provide us with a third type; and all these orders can themselves be ordered by showing the kind of relationships which exist among them, how they interact with one another on both the synchronic and the diachronic levels" (p. 312). Trinh (1989) writes: "Science is Truth, and what anthropology seeks first and foremost through its noble defense of the native's cause (whose cause? you may ask) is its own elevation to the rank of Science" (p. 57).

9. Anthropological discourse is not understandable by those who are its object, but anthropologists have generally not considered this a problem. As Malinowski (1922) writes: "Unfortunately, the native can neither get outside his tribal atmospheres and see it objectively, nor if he could, would he have intellectual and linguistic means sufficient to express it" (p. 454). Likewise, Descola (1996) writes: "The

underlying logic detected by scholarly analysis seldom rises into the conscious minds of the members of the culture that he is studying. They are no more capable of formulating it than a young child is capable of setting out the grammatical rules of a language that he has, notwithstanding, mastered" (p. 144).

10. Lévi-Strauss (1949a pp. 154–155).

11. Rosaldo (1989 p. 180). Bourdieu (1990) writes: "Undue projection of the subject onto the object is never more evident than in the case of the primitivist participation of the bewitched or mystic anthropologist, which, like populist immersion, also plays on the objective distance from the object to play the game as a game while waiting to leave it in order to tell it. This means that participant observation is, as it were, a contradiction in terms (as anyone who has tried to do it will have confirmed in practice)" (p. 34). The published translation of Bourdieu's paragraph is imprecise, and I have rectified it here; see the French original, Bourdieu (1980 p. 57) in comparison.

12. Bourdieu (1977) was the first to explain the pernicious effects of the objectivist gaze and the immobilization of time it implies. See also Bourdieu (1990 p. 26) on the limits of objectivism. Lévi-Strauss (1963a, p. 378) writes that "the anthropologist is the astronomer of the social sciences."

13. Tsing (1993) talks of "disciplinary conventions that link domination and description" (p. 32). See also Lewis (1973) and Saïd (1978). Foucault (1961) first pointed out the will to power inherent in the clinical gaze of the social sciences. For the "unbiased and supracultural language of the observer," see Bourguignon (1970, p. 185).

14. Lévi-Strauss (1991a, p. 2).

15. The word "shaman" comes from the Tungusic word *saman*, the original etymology of which may be foreign. Different authors have proposed a Chinese origin (*sha-men* = witch), a Sanskrit origin (*sramana* = buddhist monk), and a Turkish origin *(kam)*—see Eliade (1964, pp. 495–499). Lot-Falck (1963, p. 9) gives an indigenous etymology which she presents as "universally recognized nowadays": the Tungusic root *sam-*, which signifies the idea of body movement. She concludes: "All the observers of shamanism have therefore been justifiably struck by this gestural activity which gives its name to shamanism" (p. 18). However, Lot-Falck goes on to write ten years later: "The term 'shaman' was borrowed from the Tungusic *saman*, the etymology and origin of which are still doubtful" (1973, p. 3). Meanwhile Diószegi (1974, p. 638) proposes the Tungusic

verb *"sa-"* (= to know) as the origin of the word *saman*, which would therefore mean "the one who knows." Surprisingly, several authors base themselves on Lot-Falck's first text to claim that the word *saman* is etymologically linked to the idea of movement—see, for example, Hamayon (1978, p. 55), Rouget (1980, p. 187), and Chaumeil (1983, p. 10).

16. For summaries and bibliographies concerning the anthropology of shamanism in the late nineteenth and early twentieth centuries see Eliade (1964, pp. 23–32), Lewis (1971, pp. 178–184), Delaby (1976), and Mitriani (1982).

17. Devereux (1956, pp. 28–29).

18. Lévi-Strauss (1949b), published in Lévi-Strauss (1963a, pp. 197–199).

19. Lewis (1971): "The shaman is not the slave, but the master of anomaly and chaos. In rising to the challenge of the powers which rule his life and by valiantly overcoming them in this crucial initiatory rite which reimposes order on chaos and despair, man reasserts his mastery of the universe and affirms his control of destiny and fate" (pp. 188–189). Browman and Schwarz (1979): "Anthropologists use the term 'shaman' to refer to persons encountered in nonliterate cultures who are actively involved in maintaining and restoring certain types of order" (p. 6). Hamayon (1982): "On the other hand, what can distinguish the shamanic system is that it defines itself in terms of disorder, which is to be avoided, and not in terms of order, which is to be maintained" (p. 30). Hoppál (1987): "Shamans as mediators create order and reestablish balance within their groups such that their role is socially embedded in their cultures" (p. 93).

20. In his 1967 article entitled "Shamans and acute schizophrenia," Silverman writes that shamans and schizophrenics both exhibit "grossly non-reality-oriented ideation, abnormal perceptual experiences, profound emotional upheavals, and bizarre mannerisms" (p. 22). Since then, the view that shamans are mentally ill has withered, but has not entirely disappeared. Lot-Falck (1973) writes that "one can hardly contest that shamans are abnormal beings" (p. 4); Hultkrantz (1978) writes: "Our conclusion is, then, that the shaman has a hysteroid disposition which, however, does not provoke any mental disorder" (p. 26); Perrin (1992a) writes: "In other words, the first shamans would have been 'real hysterics' before the system they created became entirely accepted as a logical and formal representation, made up of elements of hysterical nature, but which are

now semi-independent of their psychological origin" (p. 122). Finally, Noll (1983) provides a demonstration of the fundamental differences between shamanism and schizophrenia.

21. Browman and Schwarz (1979, p. 7). See Halifax (1979, pp. 3–4) for a similar jack-of-all-trades definition of the shaman.

22. Taussig (1987) writes: "But what would happen if instead of this we allow the old meaning to remain in the disorder, first of the ritual, and second of the history of the wider society of which it is part? My experience with Putumayo shamans suggests that this is what they do, and that the magical power of an image like the Huitoto lies in its insistently questioning and undermining the search for order" (p. 390). Brown (1988), in discussing the "anti-structural world of the Aguaruna shaman," considers the latter's work to involve "struggle, uncertainty, ambivalence and partial revelation." According to Brown, the function of the shaman's revelations is to "shift disorder from the human body to the body politic" (pp. 115, 103, 102).

23. See Eliade (1964), p. 5 ("specializes in a trance"), pp. 96–97 ("secret language"), pp. 126ff. and 487ff. (vines, ropes, ladders), and p. 9 ("spirits from the sky").

24. See Hamayon (1990, pp. 31–32—latent mysticism), Delaby and Hamayon quoted in Chaumeil (1983, p. 16—detaching symbols from their context), Hamayon (1978, p. 55—Eliade's mysticism mutilates and distorts the facts, obliterating the sociocultural aspect of the shamanic institution and practice), and Chaumeil (1983, p. 17—the mystical dead end into which Eliade locks the phenomenon). All these references are cited by Chaumeil (1983, pp. 16–19). Taussig (1992, p. 159) calls Eliade's work "a potentially fascistic portrayal of third world healing."

25. Geertz (1966, p. 39). Furthermore, Taussig (1989, quoted in Atkinson 1992, p. 307) writes that "shamanism is . . . a made-up, modern, Western category, an artful reification of disparate practices, snatches of folklore and overarching folklorizations, residues of long-established myths intermingled with the politics of academic departments, curricula, conferences, journal juries and articles, [and] funding agencies." The first anthropologist to criticize the concept of shamanism was Van Gennep, who protested, in 1903, against the use of an obscure Siberian word to describe the beliefs and customs "of the semi-civilized the world over" (p. 52).

26. See Lévi-Strauss (1963b).

27. Luna (1986, pp. 62, 66).

1. See Swenson and Narby (1985, 1986), Narby (1986), Beauclerk, Narby, and Townsend (1988), and Narby (1989).
2. Until recently, and for unknown reasons, Spanish speakers have called the Ashaninca "Campas." The etymology of this word is doubtful. As Weiss (1969) writes: "The term 'Campa' is not a word in the Campa language" (p. 44). According to him, the word probably comes from the Quechua "tampa" ("in disorder, confused") or "ttampa" ("disheveled") (p. 61). However, there is no agreement among specialists on the word's exact etymology—see Varese (1973, pp. 139–144). Renard-Casevitz (1993) justifies her use of the word "campa" as follows: "The term campa is not appreciated as an ethnonym, though it does present a certain convenience. . . . I use campa for want of a term with a comparable reach to designate the totality of the Arawak subsets who share a notable cultural trait: the prohibition of internal war, among all except the Piro" (pp. 29, 31). In the 1980s, one of the first demands put forth by the different Ashaninca organizations was that people stop designating them by a name that they do not use in their own language.
3. See Weiss (1969, pp. 93, 96, 97–100, 201).
4. See Weiss (1969, pp. 107–109, 199–226). The quote is on page 222.
5. Weiss (1969, p. 200).
6. For a more detailed account of this experience, see Narby (1990, pp. 24–27).

4: Enigma in Rio

1. Eight indigenous land-titling projects were carried out successfully, covering a total of 2,303,617 hectares (23,000 km^2 or 5,692,237 acres). Details concerning these projects can be obtained from "Nouvelle Planète," CH-1042 Assens, Switzerland.
2. The Rio Declaration states: "Indigenous people and their communities . . . have a vital role in environmental management and development because of their knowledge and traditional practices. States should recognize and duly support their effective participation in the achievement of sustainable development" (Principle 22). The Agenda 21 underlines the importance of the territorial rights of indigenous peoples and of their self-determination in matters of development (Chapter 26). The Statement of Forest Principles points

out the importance of respecting the rights and interests of indigenous peoples and of consulting them on forestry policies (Points 2d, 5a, 13d). The Convention on Biological Diversity considers the importance of the knowledge and practices of indigenous peoples and calls for their equitable remuneration (Points 8j, 10c, 10d). The Rio conference was a spectacular turning point for indigenous rights. Just five years beforehand, the question of these rights remained largely ignored by most international organizations concerned with development or environmental matters.

3. For example, The Body Shop and Shaman Pharmaceuticals, whose vice-president declared: "Shaman [Pharmaceuticals] is committed to providing direct and immediate reciprocal benefits to indigenous people and the countries in which they live" (King 1991, p. 21).

4. These figures come from, respectively, Farnsworth (1988, p. 95), Eisner (1990, p. 198), and Elisabetsky (1991, p. 11).

5. Estimates of the number of "higher" (that is, flowering) plant species vary from 250,000 to 750,000. Wilson (1990) writes: "How much biodiversity is there in the world? The answer is remarkable: No one knows the number of species even to the nearest order of magnitude. Aided by monographs, encyclopedias, and the generous help of specialists, I recently estimated the total number of described species (those given a scientific name) to be 1.4 million, a figure perhaps accurate to within the nearest 100,000. But most biologists agree that the actual number is at least 3 million and could easily be 30 million or more. In a majority of particular groups the actual amount of diversity is still a matter of guesswork" (p. 4).

6. The Convention on Biological Diversity mentions the importance of "equitable" remuneration for indigenous knowledge, but fails to provide a mechanism to this effect. According to the Kari-Oca Declaration signed by the delegates of the World Conference of Indigenous Peoples on Territory, Environment and Development (May 1992): "The usurping of traditional medicines and knowledge from indigenous peoples should be considered a crime against peoples" (Point 99). Furthermore: "As creators and carriers of civilizations which have given and continue to share knowledge, experience and values with humanity, we require that our right to intellectual and cultural properties be guaranteed and that the mechanism for each implementation be in favor of our peoples and studied in depth and implemented. This respect must include the right over genetic re-

sources, gene banks, biotechnology and knowledge of biodiversity programs" (Point 102). See also Christensen and Narby (1992).

7. Tubocurarine is the best-known active ingredient of Amazonian curare preparations, but, as Mann (1992) points out, C-toxiferine is twenty-five times more potent. However, "both drugs have been largely superseded by other wholly synthetic neuromuscular blocking agents, such as pancuronium and atracurium. Like tubocurarine these have a rigid molecular structure with two positively charged nitrogen atoms held in a similar spatial arrangement to that found in tubocurarine. This allows them to bind to the same acetycholine receptor and thus mimic the biological activity of tubocurarine, because the distance between the two cationic centres (N^+ to N^+ distance) is approximately the same" (pp. 21–23). Concerning the initial use of curare in medicine, see Blubaugh and Linegar (1948).

8. See Schultes and Raffauf (1990, pp. 265ff. and 305ff.) for a relatively exhaustive list of the different plant species used across the Amazon Basin for the production of curare. As Bisset (1989) points out, the chemical activity of Amazonian curares is still poorly understood. Most of these muscle-paralyzing substances contain plants of the *Strychnos* or *Chondodendron* genus, or a combination of both, to which a certain number of admixtures are added, according to the recipes. The exact role of these admixtures is obscure, even though they seem to contribute to the potentiation of the main ingredients. Moreover, Manuel Córdova (in Lamb 1985) provides a first-person account of the production of curare destined for medical use, in which he repeatedly mentions the importance of avoiding "the pleasantly fragrant vapors" (p. 48)—giving the example of a German zoologist who died for lack of care (pp. 97–98). First-person accounts of curare production are rare, as curare recipes are often jealously guarded secrets.

9. See Reichel-Dolmatoff (1971, pp. 24, 37).

10. For examples of texts that illustrate the value of the botanical knowledge of Amazonian peoples with multiple references to curare, *Pilocarpus jaborandi,* and *tikiuba,* see the special issue of *Cultural Survival Quarterly* (Vol. 15, No. 3) devoted to the question of intellectual property rights of indigenous peoples, and in particular the articles by Elisabetsky (1991), Kloppenberg (1991) and King (1991). On the more general question of these rights, see Posey (1990, 1991). See Rouhi (1997) for references to *Couroupita guienensis*

and *Aristolochia*. For recent work on the unidentified plants of the indigenous pharmacopoeia, see Balick, Elisabetsky, and Laird (1996), in particular the article by Wilbert (1996), as well as Schultes and von Reis (1995).

11. See Luna (1986, p. 57).

12. Schultes and Raffauf (1992, p. 58). Davis (1996) writes: ". . . Richard Evans Schultes, the greatest ethnobotanist of all, a man whose expeditions . . . placed him in the pantheon along with Charles Darwin, Alfred Russel Wallace, Henry Bates, and his own hero, the indefatigable English botanist and explorer Richard Spruce" (p. 11). Davis's book is a treat, beautifully written and well researched.

13. Slade and Bentall (1988) write: "Indeed, taking the ordinary language words 'real' and 'imaginary' to describe public and private events respectively, it is true by definition that the act of hallucination involves mistaking the 'imaginary' for the 'real'" (p. 205). Hare (1973) writes: "Let us instead define a hallucination as a subjective sensory experience which is of morbid origin and interpreted in a morbid way" (p. 474). *Webster's Third New International Dictionary* defines *hallucination* as follows: "perception of objects with no reality; experience of sensations with no external cause usually arising from disorder of the nervous system; . . . a completely unfounded or mistaken impression or notion; Delusion."

14. According to Renck (1989), who reviewed the scientific literature on the matter, and who bases himself on Tavolga's work, there are six levels of communication: *vegetative* (the color of the flower, the texture of the fur), *tonic* (the smell of the flower, the heat of the body), *phasic* (the chameleon changes skin color, the dog pricks up its ears), *descriptive* (the dog growls), *symbolic* (some monkeys can communicate with abstract signs), and *linguistic* ("The only known example is the language articulated by man," p. 4).

5: DEFOCALIZING

1. The Young Gods, and Steve Reich.

2. See Crick (1994, pp. 24, 159) on the visual system, and more broadly Penrose (1994) and Horgan (1994) on the current limits of knowledge about consciousness.

3. Among the exceptions, Hofmann (1983, pp. 28–29) writes: "As yet we do not know the biochemical mechanisms through which LSD

exerts its psychic effects"; Grinspoon and Bakalar (1979, p. 240) write on the main effects of hallucinogens: "The only reasonably sure conclusion we can draw is that their psychedelic effects are in some way related to the neurotransmitter 5-hydroxytryptamine, also called serotonin. Not much more than that is known"; and Iversen and Iversen (1981) write: "We remain remarkably ignorant of the scientific basis for the action of any of these drugs." See the bibliographies in Hoffer and Osmond (1967) and in Slade and Bentall (1988) for an overview of the numerous studies on hallucinations and hallucinogens during the 1950s and 1960s.

4. Schultes and Hofmann (1979, p. 173).

5. Psilocybin, which is found in over a hundred mushroom species, is a close variant of dimethyltryptamine, as Schultes and Hofmann (1980) write: "Degradation studies showed psilocybin to be a 4-phosphoryloxy-N,N-dimethyltryptamine. Hydrolysis of psilocybin gives equi-molecular amounts of phosphoric acid and psilocin, which is 4-hydroxy-N,N-dimethyltryptamine" (p. 74). LSD is 100 times more active than dimethyltryptamine. See Hofmann (1983, p. 115) for the comparison between LSD and psilocybin, and Strassman et al. (1994) for an estimate of the basic dose of dimethyltryptamine.

6. Grinspoon and Bakalar (1979) write: "Used to describe the estheticized perception or fascination effect, enhanced sense of meaningfulness in familiar objects, vivid closed-eye imagery, visions in subjective space, or visual and kinesthetic distortions induced by drugs like LSD, 'hallucination' is far too crude. If hallucinations are defined by failure to test reality rather than merely as bizarre and vivid sense impressions, these drugs are rarely hallucinogenic" (pp. 6–7). However, these authors consider that the term "pseudo-hallucinogenic" is awkward, even if it describes precisely the effects of substances such as LSD and MDMA ("Ecstasy"). Slade (1976) writes: "The experience of true hallucination under mescalin and LSD-25 intoxication is probably fairly infrequent" (p. 9). For a discussion of the concept of "pseudo-hallucination," see Kräupl Taylor (1981). Regarding the evolution of the relationship between science and hallucinogens, see Lee and Shlain (1985). Finally it should be noted that the synthetic compound known as "Ecstasy" differs from the other substances mentioned here in that it appears to be neurotoxic and to destroy the brain's serotonin-producing cells (see McKenna and Peroutka 1990).

7. Besides the 72 ayahuasca-using peoples of Western Amazonia, there are those who sniff dimethyltryptamine-containing powders of vegetal origin, or who lick dimethyltryptamine-containing pastes. These pastes and powders are made from different plants (*Virola, Anadenanthera, Iryanthera,* etc.) depending on the region. Sniffing dimethyltryptamine powders also seems to have been a custom among the indigenous peoples of the Caribbean, until they were physically eliminated during the sixteenth and seventeenth centuries.

8. As I noted in Chapter 2, the exact chemical composition of ayahuasca remains a mystery. It should be pointed out that, contrary to the recent scientific studies which indicate that dimethyltryptamine is the brew's main active ingredient, ayahuasqueros consider that *Banisteriopsis caapi* (containing the beta-carbolines) is the main ingredient, and that *Psychotria viridis* (containing the dimethyltryptamine) is only the additive—see Mabit (1988) and Mabit et al. (1992). Regarding scientific research on the effects of dimethyltryptamine, the studies by Szára (1956, 1957, 1970), Sai-Halasz et al. (1958), and Kaplan et al. (1974) all consider this substance as a "psychotomimetic" or a "psychotogen," an imitator or a generator of psychosis. The study by Strassman et al. (1994) is the only one that I found with a neutral approach. However, all of these studies agree on one point: Dimethyltryptamine produces true hallucinations, in which the visions replace normal reality convincingly. As Strassman et al. (1994) write: "Reality testing was affected inasmuch as subjects were often unaware of the experimental setting, so absorbing were the phenomena" (p. 101). Finally it is worth noting that there are several interesting non-scientific studies, provided by people who have used this substance, published in Stafford (1977, pp. 283–304), as well as the writings of Terence McKenna (1991).

9. Slade and Bentall (1988) attribute the vertiginous speed of certain visual hallucinations to "the known time-distorting effects of hallucinogens" (p. 154)—but I find this explanation to be insufficient in the light of my personal experience; under the influence of ayahuasca I saw images fly past at unimaginable speed without feeling a chronological acceleration in any other domain of my internal reality. Siegel and Jarvik (1975) sum up the usual scientific theory on the internal and cerebral origin of hallucinatory images: "The notion of hallucinations consisting of complex memory imagery is neither a radical nor a new idea. It is not radical because it appeals to an intuitive sense of what is reasonable to infer. When one hallucinates

something that is not there, the stimuli being perceived (i.e., the image) must come from some source. It is not reasonable for normal man to infer that such stimuli, when auditory, are 'voices talking to me,' 'radio waves from another planet,' or clairvoyant communications with a deceased loved one. Nor is it always reasonable to infer that the stimuli, when visual, are real (e.g., 'that little green man is really there') or self-contained in a recently administered drug (e.g., 'God is in LSD'). Rather, it is more reasonable to infer that such phenomena originate in stored information in the brain, that is, memories" (p. 146).

10. In the nineteenth century, botanist Richard Spruce and geographer Manuel Villavicencio both described their personal ayahuasca experiences—see Reichel-Dolmatoff (1975, Chapter 2) for extracts of their reports. Currently, there is a range of positions within anthropology concerning the investigator's personal use of hallucinogens. Taussig (1987), who uses the Colombian term *yagé* for ayahuasca, writes: "There is no 'average' *yagé* experience; that's its whole point. Somewhere you have to take the bit between your teeth and depict *yagé* nights in terms of your own experience" (p. 406). At the other end of the spectrum, Chaumeil (1983) writes: "Moreover, I was never truly initiated into shamanic practices, which certainly gave me an external vision of the phenomenon, but which also guaranteed, on the other hand, a certain 'objectivity'" (p. 9). Strangely, even though I feel a greater affinity for Taussig's perspective—his book stimulated my thinking on how to broach the subject of Amazonian hallucinogens—I found Chaumeil's book more useful for clarifying questions of techniques and content. This seems to indicate that it is possible to be a good film critic without ever seeing a movie with one's own eyes, but by interviewing film buffs with patience and method—as Chaumeil did with Yagua shamans.

11. Harner (1968, pp. 28–29).

12. Buchillet (1982, p. 261).

13. All quotes are from Harner (1980, pp. 1–10).

14. Reichel-Dolmatoff (1981, p. 81).

15. Ibid. (p. 87).

16. Ibid. (p. 78).

17. See Chaumeil (1983, pp. 148–149) for the two quotes. The "celestial serpent" appears in the drawing entitled "Schéma 1" on the unnumbered page between pages 160 and 161.

1. Most authors report that ayahuasca is taken in complete darkness, which guarantees tranquillity to a certain extent and enhances the visions—see Kensinger (1973, p. 10), Weiss (1973, p. 43), Chaumeil (1983, p. 99), Luna (1986, p. 147), and Baer (1992, p. 87). According to Gebhart-Sayer (1986), Shipibo-Conibo shamans wait for their neighbors' hearth fires and lamps to go out before drinking ayahuasca "given that light damages their eyes during the visions" (p. 193). However, Reichel-Dolmatoff (1972, p. 100) reports that the Tukano drink ayahuasca in the light of a red torch; Luna (1986, p. 145) reports that one of his informants had occasionally participated in sessions occurring on moonlit evenings and Whitten (1976, p. 155) describes a session which took place "around a very low-burning fire."

2. Regarding the presence of bananas and fish in the ayahuasqueros' diet, see Métraux (1967, p. 84), Lamb (1971, p. 24), Reichel-Dolmatoff (1975, p. 82), Whitten (1976, p. 147), Chaumeil (1983, p. 101), Luna (1984, p. 145), and Descola (1996, p. 339). The only mention I found of the connection between this diet and neurotransmitters was in a talk by Terence McKenna (1988, Cassette 5, Side B). On the concentration of serotonin in fish and bananas, see Hoffer and Osmond (1967, p. 503). In the short term, substances such as dimethyltryptamine displace serotonin by bonding to its receptors; this causes the synaptic levels of serotonin to rise and only hinders the brain's overall production of serotonin in the long term, after repeated use; it is precisely under these circumstances that ayahuasqueros eat bananas and fish. According to Pierce and Peroutka (1989): "Biochemical studies have demonstrated the indolealkylamines [such as dimethyltryptamine and LSD] suppress 5-HT [serotonin] metabolism and decrease levels of 5-hydroxyindoleacetic acid and increase synaptosomal levels of 5-HT" (p. 120). Descola (1996) writes regarding the diet of apprentice ayahuasqueros among the Achuar: "The resulting diet is dauntingly dull, its basis being plantains (from which the pips must be removed) and boiled palm hearts, sometimes accompanied by small fish" (p. 339). He explains these "dietary prohibitions," or "taboos," as follows: "However irrational they may seem, taboos may be regarded as an effect produced by classificatory thinking. Because they draw attention to a system of concrete properties signified by a limited collec-

tion of natural species—properties that make the point that no person is exactly like any other in that the flesh of these species is proscribed for him or her personally either temporarily or permanently—taboos testify to a desire to confer order upon the chaos of the social and natural world, purely on the basis of the categories of physical experience" (p. 340).

3. Suren Erkman, personal communication, 1994.

4. The quote is from Townsley (1993, pp. 452, 456). Ayahuasqueros generally consider the mothers, or animate essences, of plants to be the sources of their knowledge. Chaumeil (1983) writes regarding Yagua shamanism: "Every initiation begins with the ingestion of decoctions made from hallucinogenic plants, or plants considered as such, which allow the novice to apprehend the invisible world and to 'see,' *renuria,* the essence of beings and things, and above all the *mothers* of the plants who are the true holders of knowledge. The importance of hallucinogens in the process of gaining access to knowledge is clearly attested here; they are the main way. It is during such sessions that the novice will contact the *mothers* who, much more than the instructor shaman, will transmit the knowledge to him" (original italics, p. 312). Regarding these *mothers,* Chaumeil writes: "Everything that is animated, *siskatia,* 'which lives,' has an essence, *hamwo,* or *mother* on which the shaman can act. On the contrary, all that is lacking one is *ne siskatia,* 'inanimate,' 'lifeless'" (p. 74). Luna (1984) writes regarding the *vegetalistas* of the city of Iquitos: "All four informants insist that the spirits of the plants taught them what they know" (p. 142). According to Reichel-Dolmatoff (1978), the Tukano acquire their artistic knowledge from the hallucinatory sphere. Gebhart-Sayer (1986, 1987) reports the same thing among the Shipibo-Conibo. Regarding the spirits, mothers, and animate essences more generally, see also Dobkin de Rios (1973), Chevalier (1982), Baer (1992), and Illius (1992).

5. Métraux (1946) writes at the beginning of his article entitled "Twin heroes in South American mythology": "A pair of brothers, generally twins, are among the most important protagonists of South American folklore. They appear as culture heroes, tricksters and transformers. The Creator or Culture Hero himself is rarely a solitary character. In many cases he has a partner who is often a powerful rival, but who may be a shadowy and insignificant personage. . . . Whenever the partner of the Culture Hero is represented as an opponent or as a mischievous or prankish character, the mythical pair

is indistinguishable from the Twin Heroes" (p. 114). Garza (1990) writes regarding Nahua and Maya shamanism: "We see the governing *nagual*, in the plastic arts of the classical period, emerging from the mouth of enormous serpents, which are magnificent, in other words plumed, and which symbolize water and the sacred vital energy" (p. 109).

6. Lévi-Strauss (1991b, p. 295).
7. See Eliade (1964, pp. 129, 275, 326, 430, 487–490). Métraux (1967) writes regarding the consecration ceremony of the young shaman among the Araucanians: "One prepares, first of all, the sacred ladder or *rewe*, which is the symbol of the profession" (p. 191).
8. As I wrote in Chapter 2, anthropologists have accused Eliade of "detaching symbols from their contexts," among other things. I must admit that I, too, had several prejudices regarding his work. The first time I read his book on shamanism and noted the repeated references to ladders, I thought Eliade simply had a folkloric obsession for the "ritual" objects of exotic cultures. I had other reasons for considering his book not to be very useful for the research I was conducting. Eliade considers "narcotic intoxication" to be a "decadence in shamanic technique" (1964, p. 401). This opinion has often been quoted over the last decades to depreciate Amazonian shamanism and its use of plant hallucinogens (which are certainly not "narcotic"). It is important to remember, however, that Eliade originally wrote his book on shamanism in 1951, before the scientific community became aware of the effects of hallucinogens. According to Furst (1994, p. 23), Eliade changed his mind toward the end of his life. The quote regarding the "Rainbow Snake" is from Eliade (1972, p. 118). Regarding crystals, he writes: "It is Ungud [the Rainbow Snake] who gives the medicine man his magic powers, symbolized by the *kimbas*, which are quartz crystals" (p. 87).
9. Campbell (1964, p. 11).
10. Campbell (1968, p. 154).
11. Chevalier and Gheerbrant (1982, pp. 867–868).
12. The quotes are from Campbell (1964, pp. 17, 9, 22). Campbell writes regarding the twin beings in the Garden of Eden: "They had been one at first, as Adam; then split in two, as Adam and Eve" (p. 29). However, "the legend of the rib is clearly a patriarchal inversion" (p. 30), as the male begets the female, which is the opposite of previous myths and of biological reality. Meanwhile, the damnation of the serpent is particularly ambiguous; Yahweh accuses it of having

shown Eve the tree that allows one to tell the difference between good and evil; how can one apply the Ten Commandments without an understanding of this difference? According to Campbell, these patriarchal inversions "address a pictorial message to the heart that exactly reverses the verbal message addressed to the brain; and this nervous discord inhabits both Christianity and Islam as well as Judaism, since they too share in the legacy of the Old Testament" (p. 17).

13. See Campbell (1964, p. 22) and Chevalier and Gheerbrant (1982, p. 872).

14. Reichel-Dolmatoff (1975, p. 165). He adds: "Now, the phenomenon of macroscopia, the illusion of perceiving objects much larger than they are, is frequent in hallucinations induced by narcotic snuff" (p. 49). This phenomenon is frequently mentioned in the hallucinogen literature. It also calls to mind *Alice's Adventures in Wonderland*, when Alice becomes extremely small after eating a piece of mushroom on which a caterpillar is smoking a hookah. Meanwhile, Descola (1996) writes regarding his personal experience with ayahuasca: "Curiously enough, these unanchored visions do not obscure the still landscape that frames them. It is rather as though I were looking at them through the lens of a microscope operating as a window of variable dimensions set in the middle of my usual and unchanged field of vision" (pp. 207–208).

15. Gebhart-Sayer (1986) writes concerning the visual music perceived by Shipibo-Conibo shamans: "This spirit [of ayahuasca] projects luminous geometric figures in front of the shaman's eyes: visions of rhythmic undulation, of perfumed and luminous ornamentation, or the rapid skimming over of the pages of a book with many motifs. The motifs appear everywhere one looks: in star formations, in a person's teeth, in the movements of his tuft of grass. As soon as the floating network touches his lips and crown, the shaman can emit melodies that correspond to the luminous vision. 'My song is the result of the motif's image,' says the shaman to describe the phenomenon, a direct transformation of the visual into the acoustic. 'I am not the one creating the song. It passes through me as if I were a radio.' The songs are heard, seen, felt and sung simultaneously by all those involved" (p. 196). The notion that ayahuasqueros learn their songs directly from the spirits is generalized. According to Townsley (1993), Yaminahua shamans "are adamant that the songs are not ultimately created or owned by them at all, but by the *yoshi* them-

selves, who 'show' or 'give' their songs, with their attendant powers, to those shamans good enough to 'receive' them. Thus, for instance, in their portrayal of the process of initiation, it is the *yoshi* who teach and bestow powers on the initiate; other shamans only facilitate the process and prepare the initiate, 'clean him out' so as to receive these spirit powers" (p. 458). Likewise, according to Luna (1984): "The spirits, who are sometimes called *doctorcitos* (little doctors) or *abuelos* (grandfathers), present themselves during the visions and during the dreams. They show how to diagnose the illness, what plants to use and how, the proper use of tobacco smoke, how to suck out the illness or restore the spirit to a patient, how the shamans defend themselves, what to eat, and, most important, they teach them *icaros*, magic songs or shamanic melodies which are the main tools of shamanic practices" (p. 142). Chaumeil (1993) talks of the extremely high-pitched sounds emitted by the spirits who communicate with Yagua shamans, more particularly of "strange melodies, both whistled and 'talked,' with a strong feminine connotation" (p. 415). Regarding the learning of songs by imitation of the spirits, see also Weiss (1973, p. 44), Chaumeil (1983, pp. 66, 219), Baer (1992, p. 91), and Townsley (1993, p. 454). See Luna (1986, pp. 104ff.) regarding the different functions of the songs (call the spirits, communicate with them, influence hallucinations, cure). See also, more generally, Lamb (1971), Siskind (1973), Dobkin de Rios and Katz (1975), Chevalier (1982), Luna and Amaringo (1991), Luna (1992), and Hill (1992). Finally, Bellier (1986) writes that among the Mai Huna of the Peruvian Amazon, "it is inconceivable to take *yagé* [ayahuasca], to penetrate the primordial world *(miña)* and to remain silent" (p. 131).

16. Luna and Amaringo (1991, pp. 31, 43). Luna writes: "I asked Pablo how he conceives and executes his paintings. He told me that he concentrates until he sees an image in his mind—a landscape, or a recollection of one of his journeys with ayahuasca—complete, with all the details. He then projects this image upon the paper or canvas. 'When this is done, the only thing I do is just add the colors.' When painting his visions he often sings or whistles some of the *icaros* he used during his time as *vegetalista*. Then the visions come again, as clear as if he were having the experience again. Once the image is fixed in his mind, he is able to work simultaneously with several paintings. He knows perfectly well where each design or color will go. In his drawings and paintings there are no corrections—in the

five years since we met he has never thrown away one single sheet of paper. Pablo believes that he acquired his ability to visualize so clearly and his knowledge about colors partly from the ayahuasca brew" (p. 29).

17. Suren Erkman, personal communication, 1994.
18. Jon Christensen, personal communication, 1994.
19. See Crick (1981, pp. 51, 52–53, 70). He also writes: "Consider a paragraph written in English. This is made from a set of about thirty symbols (the letters and punctuation marks, ignoring capitals). A typical paragraph has about as many letters as a typical protein has amino acids. Thus, a similar calculation to the one above would show that the number of different letter-sequences is correspondingly vast. There is, in fact, a vanishingly small hope of even a billion monkeys, on a billion typewriters, ever typing correctly even one sonnet of Shakespeare's during the present lifetime of the universe" (p. 52).

7: MYTHS AND MOLECULES

1. Angelika Gerhart-Sayer, personal communication, 1995.
2. The quotes about the Ouroboros are from Chevalier and Gheerbrant (1982, pp. 716, 868, 869), who also write that the dragon is "a celestial symbol, the power of life and of manifestation, it spits out the primordial waters of the Egg of the world, which makes it an image of the creating Verb." Mundkur (1983) writes in his exhaustive study of the serpent cult: "It is doubtful, however, that any serpent can or has ever been known to attempt to bite or 'swallow' its own tail" (p. 75).
3. According to Graves (1955), Typhon was "the largest monster ever born. From the thighs downward he was nothing but coiled serpents, and his arms which, when he spread them out, reached a hundred leagues in either direction, had countless serpents' heads instead of hands. His brutish ass-head touched the stars, his vast wings darkened the sun, fire flashed from his eyes, and flaming rocks hurtled from his mouth" (p. 134). Chuang-Tzu (1981) begins his book with this paragraph: "In the North Ocean there is a fish, its name is the K'un; the K'un's girth measures who knows how many thousand miles. It changes into a bird, its name is P'eng; the P'eng's back measures who knows how many thousand miles. When it puffs out its chest and flies off, its wings are like clouds hanging from the

sky. This bird when the seas are heaving travels to the South Ocean. (The South Ocean is the Lake of Heaven.) In the words of the Tall stories, 'When the P'eng travels to the South Ocean, the wake it thrashes on the water is three thousand miles long, it mounts spiralling on the whirlwind ninety thousand miles high, and is gone six months before it is out of breath'" (p. 43).

4. Laureano Ancon is quoted in Gebhart-Sayer (1987, p. 25). Eliade (1949) writes: "A limitless number of legends and myths represent Serpents or Dragons who control the clouds, live in ponds and provide the world with water" (pp. 154–155). According to Mundkur (1983): "Among the Aborigines of Australia, the most widespread of mythic beliefs has to do with a gigantic Rainbow Serpent, a primordial creature associated largely with beneficent powers of fertility and water. He (sometimes she) is also the source of magical quartz crystals known as *kimba* from which the medicine man derives his own power" (p. 58). According to Chevalier and Gheerbrant (1982): "The Underworld and the oceans, the primordial water and the deep earth form one single *materia prima,* a primordial substance, which is that of the serpent. Spirit of the primary water, it is the spirit of all waters, those of below, those that run on the surface of the earth, or those of above" (p. 869). Davis (1986) writes about Damballah, the Great Serpent of Haitian myth: "On earth, it brought forth Creation, winding its way through the molten slopes to carve rivers, which like veins became the channels through which flowed the essence of all life. In the searing heat it forged metals, and rising again into the sky it cast lightning bolts to the earth that gave birth to the sacred stones. Then it lay along the path of the sun and partook of its nature. Within its layered skin, the Serpent retained the spring of eternal life, and from the zenith it let go to the waters that filled the rivers upon which the people would nurse. As the water struck the earth, the Rainbow arose, and the Serpent took her as his wife. Their love entwined them in a cosmic helix that arched across the heavens" (p. 177). Davis (1996) discusses the cosmological notions of Kogi Indians as reported by Reichel-Dolmatoff: "In the beginning, they explained, all was darkness and water. There was no land, no sun or moon, and nothing alive. The water was the Great Mother. She was the mind within nature, the fountain of all possibilities. She was life becoming, emptiness, pure thought. She took many forms. As a maiden she sat on a black stone at the bottom of the sea. As a serpent she encircled the world. She

was the daughter of the Lord of Thunder, the Spider Woman whose web embraced the heavens. As Mother of Ice she dwelt in a black lagoon in the high Sierra; as Mother of Fire she dwells by every hearth. At the first dawning, the Great Mother began to spin her thoughts. In her serpent form she placed an egg into the void, and the egg became the universe" (p. 43)—see also Reichel-Dolmatoff (1987). Bayard (1987) writes regarding the serpent's symbolism: "Serpents, in their relationship with the depths of the primordial waters and of life, intertwine and establish the knot of life, which we find in the Osirian way in the druidic conception of the Nwyre" (p. 74).

5. Each human cell contains approximately 6 billion base pairs (= 6 × 10^9, meaning 6 followed by 9 zeros). Each base pair is 3.3 angstroms long [1 angstrom = 10^{-10} meters (m)]. Multiplying these two figures, one obtains 1.98 m in length, which is generally rounded to 2 m. Moreover, the double helix is 20 angstroms wide (20 × 10^{-10} m). By dividing the length by the width, one obtains a billion—see Calladine and Drew (1992, pp. 3, 16–17). The average little finger is more or less 1 centimeter wide; Paris and Los Angeles are separated by a distance of approximately 9,100 kilometers. This comparison is supposed to give a notion easy to visualize rather than an exact equation; in fact, the DNA contained in a human cell is 10 percent longer, relatively speaking, than a centimeter-wide finger stretching from Paris to Los Angeles. Moreover, in the wide spectrum of electromagnetic waves, human eyes perceive only a very narrow band, from 7 × 10^{-7} m (red light) to 4 × 10^{-7} m (violet light). De Duve (1984) writes: "Even with a perfect instrument, no detail smaller than about half the wavelength of the light used can be perceived, which puts the absolute limit of resolution of a microscope utilizing visible light to approximately 0, 25 μm" (p. 9); that is, 2,500 angstroms.

6. Wills (1989) writes that the nucleus of a cell "is about two millionths of the volume of a pinhead" (p. 22). Frank-Kamenetskii (1993) writes: "If we assume that the whole of DNA in a human cell is one molecule, its length L will be about 2 m. This is a million times more than the nucleus diameter" (p. 42). Moreover, according to some estimates, there are 100 thousand billion, or 10^{14}, cells in a human body—see, for example, Sagan and the Editors of the *Encyclopaedia Britannica* (1993, p. 965), Pollack (1994, p. 19), and Schiefelbein (1986, p. 40). However, there is no consensus on this figure.

Dawkins (1976, p. 22) uses 10^{15} ("a thousand million million"); Margulis and Sagan (1986, p. 67) use 10^{12}, but in the French translation of their book they write: "The human body is made up of 10^{16} (10 million billion) animal cells and 10^{17} (100 million billion) bacterial cells" (1989, p. 65). The difference between 10^{12} and 10^{16} is of the order of 10,000! To calculate the total length of the DNA in a human body, I chose the figure that seems to be the most widely used, and that is halfway between the extremes. When I write that our body contains 125 billion miles of DNA, or 200 billion kilometers, it is merely a rough estimate; the true number could be 100 times greater, or smaller. Finally, a Boeing 747 traveling for 75 years at 1,000 km/h would travel 657 million kilometers, which is 0.32 percent of 200 billion kilometers; the average distance between Saturn and the Sun is 1,427,000,000 kilometers.

7. Most cells contain between 70 and 80 percent water. According to Margulis and Sagan (1986): "The concentrations of salts in both seawater and blood are, for all practical purposes, identical. The proportions of sodium, potassium, and chloride in our tissues are intriguingly similar to those of the worldwide ocean. . . . we sweat and cry what is basically seawater" (p. 183–184). Without water, a cell cannot function; as De Duve (1984) writes: "Even the hardiest bacteria need some moisture around them. They may survive complete dryness, but only in a dormant state, with all their processes arrested until they are reawakened by water" (p. 21). On the relationship between water and the shape of the DNA double helix, see Calladine and Drew (1992), who write: "We see right away how DNA forms a spiral or helix on account of the low solubility in water of the bases" (p. 21).

8. Pollack (1994, pp. 29–30).

9. Both quotes are from Margulis and Sagan (1986, pp. 115–116, 111). On the terrestrial atmosphere before the apparition of life, see Margulis and Sagan (1986, pp. 41–43). They also write: "Barghoorn's Swaziland discovery of actual 3,400-million-year-old microbes raises a startling point: the transition from inanimate matter to bacteria took less time than the transition from bacteria to large, familiar organisms. Life has been a companion of the Earth from shortly after the planet's inception" (p. 72). The recently discovered traces of biological activity dating back 3.85 billion years consist of a reduced ratio of carbon-13 to carbon-12 in sedimentary rocks in Greenland—see Mojzsis et al. (1996) and Hayes (1996); Hayes writes: "The new finding seems to extend that record to the very bottom of

our planet's sedimentary pile, crucially altering earlier views of these oldest sediments and leaving almost no time between the end of the 'late heavy bombardment' of bodies within the inner Solar System by giant meteorites and the first appearance of life" (p. 21). Judson (1992) writes regarding nucleated cells ("eukaryotes"): "Eukaryotic cells are far larger than bacteria—proportionately as a horse to a bumblebee. They have hundreds of times more genes, and 500-fold more DNA" (p. 61).

10. Lewontin (1992) writes: "Fully 99.999 percent of all species that have ever existed are already extinct" (p. 119). For estimates regarding the current number of species, see Wilson (1990, p. 4, "most biologists agree that the actual number is at least 3 million and could easily be 30 million or more") and Pollack (1994, p. 170, "five million to fifty million"). Wilson (1992, p. 346) also writes: "Even though some 1.4 million species of organisms have been discovered (in the minimal sense of having specimens collected and formal scientific names attached), the total number alive on earth is somewhere between 10 and 100 million."

11. Wills (1991, p. 36). Regarding the direct observation of DNA's propensity to wriggle ("like small snakes slithering through mud"), see Lipkin (1994, p. 293). Dubochet (1993) writes: "It is not the enzyme that rotates along the DNA helix during transcription, but the DNA that rotates on itself, while moving like a supercoiled conveyor belt" (p. 2).

12. Regarding the "paradoxical passage," see Eliade (1964, p. 486). Regarding the serpent-dragon guarding the axis, see Eliade (1949, pp. 250–251), Chevalier and Gheerbrant (1982, p. 385), and Roe (1982, p. 118).

13. To describe DNA's form, Pollack (1994, p. 22) talks of "twisted vines"; Calladine and Drew (1992, pp. 24, 42, 123) of a "highly twisted ladder," a "spiral staircase," and a "snake"; Blocker and Salem (1994, p. 60) of a "spiral staircase"; Stocco (1994, p. 37) of a "ladder"; Frank-Kamenetskii (1993, p. 14) of a "rope ladder." The quote in the text ("like two lianas") is from Frank-Kamenetskii (1993, p. 92). Regarding the genetic nature of cancer, and the recent advances in scientific understanding of the phenomenon, see Sankarapandi (1994) and Jones (1993).

14. The quote is from Weiss (1969, p. 302). He also writes: "The Sky-Rope motif, which we have already encountered among the Campas and Machiguengas, and which we now find present among the

Piros, turns out to be quite widespread among the Tropical Forest tribes. It is reported, in one form or another, for the Cashinahua, the Marinahua, the Jívaro, the Canelo, the Quijo, the Yagua, the Witoto, a number of the Cuiana tribes (the Korobohana, Taulipang and Warrau), the Bacairi, the Umotina, the Bororo, the Mosetene, and the Tiatinagua; it is also reported for the Lengua, Mataco, Toba, and Vilela of the Chaco region. . . . Clearly equivalent to the concept of the Sky-Rope is that of the Sky-Ladder, reported for the Conibo, the Tucuna and the Shipaya, and that of the Sky-Tree, reported for the Sherente, the Cariri, the Chamacoco, the Mataco, the Mocoví, and the Toba—in each case comprehended as having once connected Earth with Sky. The distribution of this motif might be extended even further if we care to recognize as equivalent the idea of a chain of arrows to the sky, reported for the Conibo, the Shipibo, the Jívaro, the Waiwai, the Tupinamba, the Chiriguano, the Guarayú, the Cumana, and the Mataco" (p. 470). Weiss also notes: "Of particular interest is the Taulipang identification of the Sky-Rope with the same peculiarly stepped vine as that which the present author's Campa informants pointed out as their own *inkíteca*" (p. 505).

15. Bayard (1987) writes in his book on the symbolism of the caduceus: "First, one must retain the association of elements that we find in all civilizations, from India to the Mediterranean, including Egypt, Palestine and Sumerian Mesopotamia: the stone, the column, the truncated and sacred tree, with one or two entwined serpents. . . . The cult of the serpent is thus linked to the art of healing since the most ancient times" (pp. 161–163). Regarding the caduceus, Chevalier and Gheerbrandt (1982) write: "The serpent has a doubly symbolic aspect: one is beneficial, the other is evil, of which the caduceus represents, as it were, the antagonism and equilibrium; this equilibrium and polarity are above all those of the cosmic currents, which are figured more generally by the double spiral"; in Buddhist esotericism, for example, "the caduceus's staff corresponds to the axis of the world and the serpents to the Kundalini," the cosmic energy inside every being (pp. 153–155). See also Boulnois (1939) and Baudoin (1918) on the ancientness of this symbol. According to Bayard (1987), the two serpents of the caduceus, the yin-yang of the T'ai Chi, and the swastika of the Hindus all symbolize "a cosmic force, with opposed directions of rotation" (p. 134) See also Guénon (1962, p. 153) on the equivalence of the caduceus and the yin-yang.

16. There is a certain confusion surrounding the origin of the caduceus as the symbol of Western medicine. To start with, in Greek mythology, the caduceus's staff is the symbol of Hermes, who is, according to Campbell (1959), "the archetypal trickster god of the ancient world . . . Hermes, too, is androgyne, as one should know from the sign of his staff" (p. 417). Campbell (1964) adds that Hermes is the "guide of souls to the underworld, the patron, also, of rebirth and lord of the knowledges beyond death, which may be known to his initiates even in life" (p. 162). Hermes's staff is topped by two wings and is thus a variant on the theme of the plumed serpent. However, Hermes's staff has mainly been interpreted as a peace symbol, devoid of medical significance. The official medical caduceus is considered to belong to Aesculapius, who was said to be a real-life healer practicing around 1200 B.C., and who only became the Greek god of healing much later. Around the 5th century B.C., rationalism and patriarchy were being set up and myths were modified: Zeus, who was at first represented as a serpent, defeats the serpent-monster Typhon with the help of his daughter Athene ("Reason"), thereby guaranteeing the reign of the patriarchal gods of Olympus; concomitantly, he brings Aesculapius back to life (having previously killed him with a lightning bolt) and gives him a staff with a *single* serpent wrapped around it. According to the *Encyclopaedia Britannica*, Aesculapius's staff "is the only true symbol of medicine. The caduceus with its winged staff and intertwined serpents, frequently used as a medical symbol, is without medical relevance since it represents the magic wand of Hermes, or Mercury, the messenger of the gods and the patron of trade" (vol. 1, p. 619). To make things more complicated, the caduceus symbol, sometimes with one snake, sometimes two, has been taken up again in the twentieth century for unclear reasons. For instance, in 1902, the medical department of the United States Army adopted Hermes's staff as its symbol—while the American Medical Association took Aesculapius's staff shortly thereafter (see Friedlander 1992, pp. 127ff., 146ff.). The caduceus formed by the cup and the serpent became the official symbol of French pharmacies only in 1942 (see Burnand 1991, p. 7). The pharmacists with whom I talked invariably said that the serpent was linked to their profession "because of the venom"—for which pharmacies have antidotes.

17. Métraux (1967, pp. 191, 85, 83, 95).

18. There are many different translations of Heraclitus's fragmented work. I rely mainly on Kahn (1979). The fragment that I quote is:

"The lord whose oracle is in Delphi neither declares nor conceals, but gives a sign" (p. 43). The town of Delphi was originally called Pytho. The oracle in Delphi first belonged to the earth goddess Gaia and was defended by her child, the serpent Python. Later, Apollo slew Python and appropriated the oracle.

19. See Eliade (1964, pp. 96ff.) on the secret language of shamans. Why has there not been more interest in this language of spirits, which is reported around the world? I believe that one of the reasons is that most anthropologists do not believe that spirits *really* exist, so they cannot take them seriously. As Colchester (1982) writes in his study of the cosmovision of the Sanema in the Venezuelan Amazon: "We can only designate this spiritual realm a 'metaphoric' one, because we do not believe in its reality. Our effective understanding of Sanema phenomenology founders on this lack of belief" (p. 131). Unfortunately, Colchester's honesty is not typical.

20. The six quotes are from Townsley (1993, pp. 459–460, 453, 465). Townsley is not the only anthropologist to report the existence of a highly metaphoric shamanic language. Siskind (1973, p. 31), regarding the songs of Sharanahua ayahuasqueros, writes: "These songs are sung in an esoteric form of language, difficult to understand, and filled with metaphors." See also Colchester (1982, p. 142) on the "poetic licence" used by Sanema shamans in their songs, and Chaumeil (1993, p. 415) on the "archaic language which is incomprehensible to most" and which is used by Yagua ayahuasqueros.

21. The double helix wraps around itself completely every 10 base pairs. As there are 6 billion base pairs in a human cell, the latter's DNA wraps around itself approximately 600 million times.

22. The estimate of 97 percent of non-coding passages in the human genome is the most frequent—see, for example, Nowak (1994, p. 608) or Flam (1994, p. 1320); but Calladine and Drew (1992) consider that only 1 percent of the human genome codes for the construction of proteins (p. 14), and Blocker and Salem (1994) write: "Currently, it is generally considered that only 10% of the human genome, at most, codes for proteins; . . . No precise function has yet been found for the remaining 90% of our DNA, and it is not even certain that one will be found: it could possibly be mere 'scrap'" (p. 127). Regarding palindromes, Frank-Kamenetskii (1993) writes: "Palindromes are frequently encountered in DNA texts. Since DNA consists of two strands (i.e., as if they were two parallel texts), its palindromes may be of two types. Such palindromes in an ordinary,

single text are called 'mirrorlike.' But more frequently to be met in DNA are palindromes that read alike along either strand in the direction determined by the chemical structure oi DNA" (p. 106). The expression "junk DNA," meanwhile, was first coined by Orgel and Crick (1980) in an article entitled "Selfish DNA: The ultimate parasite," where they write: "In summary, then, there is a large amount of evidence which suggests, but does not prove, that much DNA in higher organisms is little better than junk. We shall presume, for the rest of this article, that this hypothesis is true" (pp. 604–605). See also Dawkins (1982, pp. 156ff.).

23. Calladine and Drew (1992, p. 14). Wills (1991, p. 94) estimates that there are between 30,000 and 50,000 "ACACACACACAC . . ." passages in the human genome. Nowak (1994, p. 609) estimates that the "Alu" sequence (which is 300 bases long) is repeated half a million times in the human genome. According to Watson et al. (1987, p. 668), there are several sorts of "Alu" sequences amounting to a total of a million. Jones (1993, p. 69) considers that approximately a third of the human genome is made up of repeat sequences.

24. Among the 64 words of the genetic code, only "UGG" has no synonym; it is the only word signifying the amino acid tryptophan. (The words of the genetic code are written in RNA, rather than DNA, with a U instead of a T.) The 63 other words all have at least one synonym. For instance, there are no fewer than six words for arginine: "CGU," "CGC," "CGA," "CGG," "AGA," "AGG." Moreover, two words have a double meaning: "AUG" and "GUG," which correspond respectively to amino acids methionine and valine, can also signify to the transcription enzyme where to start transcribing the text ("start"). Lewontin (1992) writes about this ambiguity: "Unfortunately, we do not know how the cell decides among the possible interpretations" (p. 67). Moreover, Watson et al. (1987) write: "Many amino acids are specified by more than one codon, a phenomenon called *degeneracy*" (p. 437, original italics). Trémolières (1994) writes: "The code is considered to be degenerate. The word is perhaps badly chosen; let us say that we are dealing with a language that has many synonyms" (p. 97).

25. Editing enzymes are called "snurps" (small nuclear ribonucleoproteins). Regarding the editing of the genetic message, Frank-Kamenetskii (1993) writes: "But what tells the enzyme how to cleave the molecule correctly and how to splice together the resulting RNA fragments? And how do in-between spaces get

dropped out in the process? The inner workings of such cutting and assembling are far from simple, for if an enzyme just cuts RNA into pieces, Brownian motion will scatter them around, with no hope for Humpty-Dumpty being put back together again" (p. 79). Blocker and Salem (1994) write: "The role of introns is extremely mysterious. Strangely, they are copied during the first stage of transcription only to end up not being transformed into 'messages.' Indeed, 'pre'-messenger RNA contains the entire gene, introns and exons. Then, still within the nucleus, a complicated mechanism takes out, or edits out, the introns. . . . Furthermore, the editing of a gene can occur in several different ways, from one time to another, often to respond to the particular demands of a given cell type. This means that this 'choice in editing' is probably strictly regulated inside each type of cell, but the way in which this regulation is realized remains almost entirely unknown" (p. 128). The alternation of exons and introns within genes is the province of "higher" organisms—in chickens, for instance, the gene corresponding to the instructions to build collagen contains fifty exons (see Watson et al. 1987, p. 629); in comparison, bacterial DNA contains practically no introns. For genes that contain up to 98 percent introns, see Wills (1991, p. 112).

26. Most estimates consider that the human genome contains 100 thousand genes. But Pollack (1994) writes: "If larger human chromosomes carry as many surprises [as yeast's], we can expect to find we are carrying, not the current estimate of one hundred thousand genes, but at least four hundred thousand genes, the majority of them unexpected and unknown" (p. 92). Meanwhile, Wade (1995b) reports on the rapid gains on the sequencing of the human genome ("which may be 99% done by 2002").

27. For the translation of these signs, see Gardiner (1950, pp. 33, 122, 457, 490, 525) and Jacq (1994, pp. 45, 204).

8: THROUGH THE EYES OF AN ANT

1. Jones (1993) writes: "A useless but amusing fact is that if all the DNA in all the cells in a single human being were stretched out it would reach to the moon and back eight thousand times" (p. 5). This calculation is based on an estimate of 3×10^{12} cells in a human body, which is 33 times smaller than the usual estimate of 10^{14} (which I use to obtain 125 billion miles of DNA in a human body).

As I explained in a note to Chapter 7, this estimate varies considerably from one specialist to another.

2. Margulis and Sagan (1986) write: "In their first two billion years on Earth, prokaryotes continuously transformed the Earth's surface and atmosphere. They invented all of life's essential, miniaturized chemical systems—achievements that so far humanity has not approached. This ancient high *bio*technology led to the development of fermentation, photosynthesis, oxygen breathing, and the removal of nitrogen gas from the air" (original italics, p. 17). Wills (1991) writes: "So the DNA molecules themselves pack over a hundred trillion times as much information by volume as our most sophisticated information storage devices" (p. 103). Pollack (1994) writes: "The second strand [of the DNA molecule] is the minimum imaginable amount of extra-molecular baggage necessary to make either strand's information self-replicating" (p. 28).

3. Luna and Amaringo (1991, pp. 33–34).

4. For the details regarding the visual system, see Ho and Popp (1993, p. 185) and Wesson (1991, p. 61).

5. See Weiss (1969), pp. 108, 202 (Avíreri, "the Great Transformer"), p. 212 ("Avíreri creates the seasons), and more generally pp. 199–226. Regarding the universality of the trickster-transformer in creation myths, Radin writes: "In the entire world there is no myth as widespread as the 'Trickster myth' that we will deal with here. There are few myths about which we can so confidently say that they belong to humanity's most ancient modes of expression; few other myths have kept their original content in such an unchanged way. The Trickster myth exists in a clearly recognizable form among the most primitive peoples as well as more evolved ones; we find it among the Ancient Greeks, the Chinese, the Japanese and in the Semitic world. . . . Though it is always linked to other myths and though it is markedly reconstructed and retold in a new form, the fundamental action seems always to have prevailed over the others" (in Jung, Kerényi, and Radin 1958, p. 7).

6. Stocco (1994, p. 38).

7. Harner (1973) writes: "Both Jívaro and Conibo-Shipibo Indians who had seen motion pictures told me that the ayahuasca experiences were comparable to the viewing of films, and my own experience was corroboratory" (p. 173).

8. In an article entitled "Evidence of photon emission from DNA in living systems," Rattemeyer et al. (1981) write: "Probably, DNA is

the most important source of 'ultra-weak' photon emission (or electromagnetic radiation) from living cells" (p. 572). On DNA's trapping and transfer of electrons, see, for example, Murphy et al. (1993), Beach et al. (1994), Clery (1995), and Hall et al. (1996); Hall et al. write: "Although the reaction we have described involves long-range photoinduced electron transfer, the precise mechanism for this DNA-mediated charge transfer is not yet known" (p. 735).

9. Wilson (1992) writes: "The black earth is alive with a riot of algae, fungi, nematodes, mites, springtails, enchytraeid worms, thousands of species of bacteria. The handful may be only a tiny fragment of one ecosystem, but because of the genetic codes of its residents it holds more order than can be found on the surfaces of all the planets combined" (p. 345). See also Wilson (1984, p. 16).

10. Margulis and Sagan (1986) write: "As soon as there were significant quantities of oxygen in the air an ozone shield built up. It formed in the stratosphere, floating on top of the rest of the air. This layer of three-atom oxygen molecules put a final stop to the abiotic synthesis of organic compounds by screening out the high-energy ultra-violet rays" (p. 112). Meanwhile, the depth of the layer of microbial life on the planet is only beginning to be investigated—see Broad (1994). Frederickson and Onstott (1996) write in their article "Microbes deep inside the earth" that they have found bacteria "from depths extending to 2.8 kilometers (1.7 miles) below the surface" (p. 45). Regarding the presence of cell-based life in the air we breathe, Krajick (1997) writes: "A cubic yard of the atmosphere can contain hundreds of thousands of bacteria, viruses, fungal spores, pollen grains, lichens, algae, and protozoa" (p. 67).

11. Quoted in Gebhard-Sayer (1987, p. 25).

12. Harner (1973) writes: "The shamans under the influence of ayahuasca see snakes apparently at least as often as any other single class of beings" (p. 161). Harner cites visions of snakes among the Jívaro, Amahuaca, Tukano, Siona, Piro, and Ixiamas Chama. According to Schultes and Hofmann (1979): "Ingestion of Ayahuasca usually induces nausea, dizziness, vomiting, and leads to either an euphoric or an aggressive state. Frequently the Indian sees overpowering attacks of huge snakes or jaguars. These animals often humiliate him because he is a mere man" (p. 121).

13. In a groundbreaking and fascinating work, Reichel-Dolmatoff (1978) gave color crayons to Desana-Tukano shamans and asked them to draw their visions; there are a good number of serpents in

these drawings—see drawings, I, IV, V, VI, VII, XVIII, XXI, XXIII, XXVI, XXVII, XXIX, XXXI, and XXXII; the latter shows two pairs of serpents wrapping around each other in spirals and, to their right, a yellow double helix; according to the caption: "This design represents four 'yagé snakes' (*gahpí piró*) that are seen after one or two cups of yagé and are in the act of climbing up the house-posts and winding around the rafters. The other, irregular, lines represent luminous sensations in the form of yellow flashes" (p. 112). Dobkin de Rios (1974) writes about the inhabitants of Iquitos who consult ayahuasqueros: "Informants repeatedly told of the boa appearing before them while under the effects of ayahuasca. However, despite the negative implications of a large, fearsome creature, this shared vision was believed to be an omen of future healing" (p. 16). See also Dobkin de Rios (1972, pp. 118–120). William Burroughs and Allen Ginsberg (1963) were among the first to write about ayahuasca; Ginsberg describes his visions: "And then the whole fucking Cosmos broke loose around me, I think the strongest and worst I've ever had it nearly . . . —First I began to realize my worry about the mosquitoes or vomiting was silly as there was the great stake of life and Death—I felt faced by Death, my skull in my beard on pallet on porch rolling back and forth and settling finally as if in reproduction of the last physical move I make before settling into real death—got nauseous, rushed out and began vomiting, all covered with snakes, like a Snake Seraph, colored serpents in aureole around my body, I felt like a snake vomiting out the universe—or a Jivaro in headdress with fangs vomiting up in realization of the Murder of the Universe—my death to come—everyone's death to come—all unready—I unready . . ." (pp. 51–52). The Cashinahua talk also of brightly colored and large snakes (see Kensinger 1973, p. 9), as does ayahuasquero Manuel Córdoba-Rios (see Lamb 1971, p. 38). Anthropologist Michael Taussig (1987) writes about his personal experience with ayahuasca: "My body is distorting and I'm very frightened, limbs stretch and become detached, my body no longer belongs to me, then it does. I am an octopus, I condense into smallness. The candlelight creates shapes of a new world, animal forms and menacing. . . . Self-hate and paranoia is stimulated by horrible animals—pigs with queer snouts, slithering snakes gliding across one another, rodents with fish-fin wings. I am outside trying to vomit; the stars and the wind above, and the corral for support. It's full of animals; moving" (p. 141). Some anthropologists drink ayahuasca without seeing snakes; Philippe Descola (1996) writes about his experience with the Achuar Jivaros: "It seems likely that the strange beings, monstrous spirits and animals in a perpetual state of metamorphosis that throng their visions—but have not yet visited me—appear to them like a succession of temporarily coagulated forms against a moving background composed of the geometric patterns whose strange beauty I am now experiencing" (p. 208)—though barely a page previous to this he also writes: "Animal forms of unrecognized species display their metamorphoses and transformations before my eyes: the water-marked skin of the anaconda merges into tortoise-shell scales that elongate into the stripes of an armadillo, then reshape into the crest of an iguana against the intense blue of the wings of a *Morpho* butterfly, then stretch into black stripes which immediately fragment into a constellation of haloes standing out against the silky fur of some large cat" (p. 207). Some people hallucinate with greater difficulty than others; the dose of the hallucinogen also plays a role; this may have influenced Descola's experiences based on "half a coffee-cupful" of ayahuasca (p. 206). According to Reichel-Dolmatoff (1975), the Desana-Tukano people can glance at a drawing of hallucinations and estimate almost exactly how many cups of ayahuasca the artist had consumed: "'This is what one sees after two cups,' they would say; or 'This one can see after six cups'" (p. 173).

14. Of the 48 paintings by Pablo Amaringo in *Ayahuasca visions* (Luna and Amaringo 1991), only three do not have serpents (nos. 1, 6, and 28). The 45 other pictures are filled with fluorescent snakes, often exceptionally large, and rather frightening. Amaringo comments on painting no. 3, called *Ayahuasca and chacruna:* "This painting represents the two plants necessary in preparing the ayahuasca brew. Out of the ayahuasca vine comes a black snake with yellow, orange and blue spots, surrounded by a yellow aura. There is also another snake, the *chacruna* snake, of bright and luminous colors. From its mouth comes a violet radiation surrounded by blue rays. The *chacruna* snake penetrates the ayahuasca snake, producing the visionary effect of these two magic plants" (p. 52). Luna writes: "By far the most conspicuous motif in Pablo's visions is the snake, which, together with the jaguar, is in turn the most commonly reported vision under the effects of ayahuasca by all tribes" (pp. 41–42). Finally, the snakes shaped like hammocks shown in painting no. 19 correspond exactly to the use of the word "hammock" to signify "anaconda" in

the twisted language of Yaminahua ayahuasqueros (see Townsley 1993, p. 459); the Yaminahua live hundreds of miles from Pucallpa, where Pablo Amaringo lives.

15. Eliade (1964, p. 497).

16. Kekulé describes his dream: "I turned the chair to the fireplace and sank into a half sleep. The atoms flittered before my eyes. Long rows, variously, more closely, united; all in movement wriggling and turning like snakes. And see, what was that? One of the snakes seized its own tail and the image whirled scornfully before my eyes. As though from a flash of lightning I awoke; I occupied the rest of the night in working out the consequences of the hypothesis" (quoted in Beveridge 1950, p. 56). The commentator I quote is Thuillier (1986, p. 386). The quote on the universality of snake dreams is from Wilson (1992, p. 349).

17. Mundkur (1983, p. 6, 8). Wilson (1984), who cites Mundkur's study, formulates the fear-of-venom theory as follows: "What is there in snakes anyway that makes them so repellent and fascinating? The answer in retrospect is deceptively simple: their ability to remain hidden, the power in their sinuous limbless bodies, and the threat from venom injected hypodermically through sharp hollow teeth. It pays in elementary survival to be interested in snakes and to respond emotionally to their generalized image, to go beyond ordinary caution and fear. The rule built into the brain in the form of a learning bias is: become alert quickly to any object with a serpentine gestalt. *Overlearn* this particular response in order to keep safe" (original italics, pp. 92–93).

18. Drummond (1981), one of the rare critics of Mundkur's theory, writes: "Mundkur finds that the relevant empirical feature is its venom: 'The serpent, in my view, has provoked veneration primarily through the power of its venom.' In making this generalization, he apparently forgets the several examples of venerated but nonvenomous serpents (i.e., boas and pythons) cited in his useful survey of the 'serpent cult.' Indeed, it would be difficult to make sense of 'The Serpent's Children' and other Amazonian anaconda myths in an ethnographic context where the fer-de-lance and bushmaster are an everyday threat to life" (p. 643). Meanwhile, Eliade (1964) writes about the costume of the Altaic shaman: "A quantity of ribbons and kerchiefs sewn to its frock represent snakes, some of them being shaped into snakes' heads with two eyes and open jaws. The tails of the larger snakes are forked and sometimes three snakes have only

one head. It is said that a wealthy shaman should have 1,070 snakes on his costume" (p. 152).

1. Weiss (1969) writes: "The Campas believe that the inability of the human eye to see the good spirits in their true form can be overcome by the continual ingestion of narcotics, especially tobacco and ayahuasca, a process that in time and with perseverance will improve the eyesight to the point where the good spirits can be seen for what they are" (p. 96). Sullivan (1988) writes in his comparative work on South American religions: "Tobacco smoke is a prime object of the craving of helper spirits, since they no longer possess fire as human beings do" (p. 653). Wilbert (1987, p. 174) lists fifteen Amazonian peoples who explicitly consider tobacco a food for the spirits; I will not repeat his work here, but will simply add to his list the Yagua, who also consider tobacco "a food for the spirits in general" (Chaumeil 1983, p. 110).

2. Wilbert (1987) writes: "In any case, tobacco craving is regarded as symptomatic of the hunger sensation of Supernaturals and is transferred from the tobacco-using practitioner to the spirit world at large. Lacking tobacco of their own, the Supernaturals are irresistibly attracted to man not just, let us say, because they enjoy the fragrance of tobacco smoke or the aroma of tobacco juice, but more basically to eat and to survive. Unfortunately, a scrutiny of the ethnographic literature gives the impression that had the idea been less exotic for Western observers or had investigators succeeded in penetrating indigenous ideology more deeply than they ordinarily did, we might have learned more often about this existential reason, as it were, behind the spirits' predilection for tobacco. Scanty as the ethnographic record may be, tobacco as spirit food, nevertheless, has been documented for a good number of societies in lowland South America, which are widely spread and numerous enough to suggest that the concept is of long standing on the subcontinent" (pp. 173–174).

3. In a human brain there are tens of billions of neurons, and they are of several sorts. Each neuron is equipped with approximately a thousand synapses, which are junction sites connecting the cells to each other. Each synapse has ten million or so receptors. The number of

neurons is frequently estimated at ten billion—see, for instance, Snyder (1986, p. 4), but Changeux (1983, p. 231) talks of "tens of billions," Wesson (1991, p. 142) puts the figure at "100 billion or so," and Johnson (1994, p. E5) proposes a bracket from "100 billion to a trillion." Sackmann (quoted in Bass 1994, p. 164) estimates the number of receptors at each synapse at "about ten million." There are approximately 50 known neurotransmitters, and a given cell can have different receptors for several of these (see Smith 1994). The nicotine and acetylcholine molecules have different shapes, but the receptor cannot tell them apart because they have the same size (10 angstroms) and the distribution of their electrical charges is similar (see Smith 1994, p. 37). Wilbert (1987) writes: "This simulation capability of nicotine has been likened to the function of a skeleton key inasmuch as it fits and opens, so to speak, all cholinergic locks of postsynaptic receptors in the body" (p. 147).

4. See the article by Changeux (1993) for a clearly illustrated presentation of nicotinic receptors. The central role played by calcium ions in the activation of DNA transcription is discussed by Farin et al. (1990), Wan et al. (1991), and Evinger et al. (1994). Concerning the activation of DNA transcription by nicotine, see also Koistinaho et al. (1993), Mitchell et al. (1993), and Pang et al. (1993). Concerning nicotine's activation of genes corresponding to the proteins that make up nicotinic receptors, see Cimino et al. (1992); the latter note, however, that most studies of nicotinic receptors have been conducted on rats, and that recent research on monkeys reveals great differences from one species to another. The rat has nicotinic receptors in its cortex, which is not the case for the monkey; the precise distribution of these receptors in the human brain is still poorly understood: "It is difficult to perform such studies in human brain since the tissue can only be obtained a long time after death and it is difficult to obtain normal young brain. For these reasons, we undertook a preliminary study on nicotinic receptor distribution in monkey brain, whose CNS [central nervous system] organization is more similar to the human CNS organization than that of the rat or chick" (p. 81). Concerning the still poorly understood cascade of reactions set off by nicotine inside the nerve cell, see Evinger et al. (1994), as well as Pang et al. (1993), who note in passing: "The mechanisms with which nicotine . . . leads to repeated self-administrative behaviour are poorly understood" (p. 162).

5. The *Nicotiana rustica* species used by shamans contains up to 18 percent nicotine (Wilbert 1987, pp. 134–136), whereas the Virginia-type tobacco leaves contain from 0.5 to 1 percent nicotine in Europe and occasionally reach 2 percent in the United States (according to the Centre for Tobacco Research, Payerne, Switzerland, personal communication, 1995). Some forms of contemporary Amazonian shamanism use cigarettes, as in the case I described in Chapter 3. However, the influence of the use of an adulterated product on the efficacy of the cure has not yet been studied. Moreover, according to the *Edict on foodstuffs* published by the Federal Chancellery of Switzerland (1991), producers are allowed to add a series of substances to tobacco "that will not exceed twenty-five percent [of the final dry product] for cigarettes, cigars and similar smoking articles and thirty percent for cut or rolled tobacco" (p. 196). These additives are divided into five categories, including moistening agents, preservatives, and flavor enhancers. The fourth category reads as follows: "d. Products for ash bleaching and combustion accelerators: aluminum hydroxide, aluminum oxide, aluminum and silicium heteroxides, aluminum sulphate, alum, silicic acid, talc, titanium dioxide, magnesium oxide, potassium nitrate, carbonic, acetic, malic, citric, tartaric, lactic and formic acids, and their components of potassium, sodium, calcium and magnesium, as well as ammonium, potassium, calcium, magnesium and sodium phosphates." The fifth category reads: "e. Adhesives: the gelling and thickening agents of the Edict of the 31st of October 1979 on additives as well as pure lac, collodion, cellulose, ethyl-cellulose, acetyl-cellulose, hydroxy-ethyl-cellulose, hydroxy-propyl-methyl-cellulose, hydroxy-ethyl-methyl-cellulose, polyvinyl acetate and glyoxal" (pp. 196–197). Unfortunately, it is not possible to obtain from the cigarette manufacturers the precise list of additives for each brand, given that the recipes for this "foodstuff" are jealously guarded.

6. Cigarettes emit 4,000 toxic substances, according to (Switzerland's) Federal Office of Public Health (1994, p. 1). Klaassen and Wong (1993) write in their article on radiation in the *Encyclopaedia Britannica:* "The largest nonoccupational radiation sources are tobacco smoke for smokers and indoor radon gas for the nonsmoking population" (vol. 25, p. 925). Martell (1982) writes in a letter published in the *New England Journal of Medicine:* "Indoor radon decay products that pass from room air through burning cigarettes into mainstream smoke are present in large, insoluble smoke particles that are selec-

tively deposited at bifurcations. Thus, the smoker receives alpha radiation at bronchial bifurcations from three sources: from indoor radon progeny inhaled between cigarettes, from ^{214}Po [polonium-214] in mainstream smoke particles, and from ^{210}Po [polonium-210] that grows into ^{210}Pb [lead-210]–enriched particles that persist at bifurcations. I estimate that the cumulative alpha dose at the bifurcations of smokers who die of lung cancer is about 80 rad (1600 rem)—a dose sufficient to induce malignant transformation by alpha interactions with basal cells" (p. 310). Evans (1993) writes in an article entitled "Cigarette smoke = radiation hazard": "In 1 year, a smoker of 1 to 2 packs per day will irradiate portions of his or her bronchial epithelium with about 8 to 9 rem. This dose can be contrasted with that from a standard chest x-ray film of about 0.03 rem. Thus, the average smoker absorbs the equivalent of the dosages from 250 to 300 chest x-ray films per year" (p. 464). Strangely enough, the radioactivity of cigarette smoke is rarely mentioned in the majority of the articles dealing with the toxicity of this product. Abelin (1993), who provides a list of the different forms of cancer provoked by cigarettes, also notes that low-tar cigarettes have a lower risk factor than normal cigarettes. However, "up until now, a lowering of the risk of heart attacks or chronic lung diseases among smokers of 'light' cigarettes has not been noticed" (pp. 15–16).

7. Weiss (1969, p. 62) notes two literal translations for *sheripiári:* "he who uses tobacco" and "he who is transfigured by tobacco." Elick (1969, pp. 203–204) suggests the word combines *sheri* ("tobacco") and *piai* ("a rather common designation for the shaman in northern South America"). Baer (1992) translates the Matsigenka word *seripi'gari* as "he who is intoxicated by tobacco"—the Matsigenka being the Ashaninca's immediate neighbors. In any case, the word means "healer" and contains the root *sheri* (or *seri*), "tobacco."

8. Johannes Wilbert, personal communication, 1994.

9. That the otherwise infallible Schultes and Hofmann omitted tobacco from their classic *Plants of the gods: Origins of hallucinogenic use* (1979) is an indication of the degree to which Western science has underestimated it. Wilbert, who has led a long and solitary campaign for the recognition of tobacco's importance in shamanism, wrote in 1972: "Tobacco *(Nicotiana spp.)* is not generally considered to be a hallucinogen. Yet like the sacred mushrooms, peyote, morning glories, *Datura,* ayahuasca, the psychotomimetic snuffs, and a whole series of other New World hallucinogens, tobacco has long

been known to play a central role in North and South American shamanism, both in the achievement of shamanistic trance states and in purification and supernatural curing. Even if it is not one of the 'true' hallucinogens from the botanist's or pharmacologist's point of view, tobacco is often conceptually and functionally indistinguishable from them" (p. 55).

10. The interaction of specific snake venoms with the different nicotinic receptors varies. Deneris et al. (1991) show that certain nicotinic receptors are sensitive to given snake toxins, but not to others, and that there is even a subclass of nicotinic receptors that is insensitive to all snake venoms. See Alberts et al. (1990, pp. 319–320) for an explanation of the central role played by nicotinic receptors in the history of ion channels and by the venom of certain snakes in their identification. Changeux (1993) provides a detailed historical outline of the evolution of the research conducted on the acetylcholine receptor, where he explains the successive stages covered by scientists and the role played by nicotine, curare, and the snake venom α-bungarotoxin. He also explains the importance of the development, in the 1980s, of new techniques which allow the determination of the exact sequence of amino acids making up the proteins that constitute the receptors.

11. Of course, the legislation on controlled substances varies from one country to another, but legislation in the United States seems to serve as a model for many other Western countries. For an exhaustive survey of American legislation on controlled substances, see Shulgin (1992). Moreover, Strassman (1991) discusses in detail the labyrinth of bureaucratic, and sometimes Orwellian, obstacles he had to surmount to obtain N,N-dimethyltryptamine and to administer it to human beings in the framework of a scientific investigation.

12. According to Strassman and Qualls (1994): "The group was high functioning, with only one subject not being a professional or student in a professional training program" (p. 86). According to Strassman et al. (1994): "Our description of subjective effects of DMT [dimethyltryptamine] used reports obtained by experienced hallucinogen users who were well prepared for the effects of the drug. In addition, these subjects . . . found hallucinogens highly desirable. Thus, our sample differed from those used to characterize hallucinogens' effects in previous studies" (p. 105). As I mentioned in Note 8 to Chapter 5, the studies by Szára (1956, 1957, 1970), Sai-

Halasz et al. (1958), and Kaplan et al. (1974) all consider dimethyl-tryptamine as a "psychotomimetic" or a "psychotogen." Concerning the use of prisoners to test this substance, see, for example, Rosenberg et al. (1963), whose article starts with the following sentence: "Five former opiate addicts who were serving sentences for violation of United States narcotic laws volunteered for this experiment" (p. 39). Leary (1966) describes his visions in a scientific and personal study of the effects of dimethyltryptamine: "A serpent began to writhe up and through the soft, warm silt . . . tiny, the size of a virus . . . growing . . . now belts of serpent skin, mosaic-jeweled, rhythmically jerking, snake-wise forward . . . now circling globe, squeezing green salt oceans and jagged brown-shale mountains with constrictor grasp . . . serpent flowing blindly, now a billion mile endless electric-cord vertebrated writhing cobra singing Hindu flute-song" (p. 93).

13. Strassman et al. (1994, p. 100).
14. Two articles published in the late 1980s (McKenna et al. 1989 and Pierce and Peroutka 1989) demonstrate that different hallucinogens activate distinct serotonin receptor subtypes. Deliganis et al. (1991) went on to show that dimethyltryptamine stimulates serotonin receptor "1A" while blocking serotonin receptor "2." According to Van de Kar (1991): "Furthermore, an understanding of the 5-HT [serotonin] receptor sub-types has led to a reevaluation of old data on the neuroendocrine effects of 5-HT agonists and antagonists" (p. 292). It had often been claimed throughout the 1980s that hallucinogens activated a single receptor (see Glennon et al. 1984). So far the precise serotonin receptors stimulated by psilocybin have not been determined.
15. According to Van de Kar (1991), serotonin receptor "3" is an ion channel, while the remaining six receptors (1a, 1b, 1c, 1d, 2, and 4) are membrane-spanning antennae. Recent research subdivides these seven serotonin receptors into fifteen subcategories—see Thiébot and Hamon (1996).
16. Pitt et al. (1994) write in their article on the stimulation of DNA by serotonin: "Thus it is apparent that a novel intracellular signalling pathway contributes to the increase in DNA synthesis caused by 5-HT [serotonin] in smooth muscle and other cells in culture" (p. 185).
17. Kato et al. (1970) administered four to eleven LSD injections to four pregnant monkeys in their third or fourth month of pregnancy.

The total amount of these doses varied from 875 micrograms/kg to 9,000 micrograms/kg; the average total dose being 4,937 micrograms/kg. An average dose for a human being is estimated at 1.5 micrograms/kg (about 100 micrograms for a person weighing 70 kg or 154 pounds). Thus, the average total dose inflicted on these monkeys was 3,000 times greater than the normal quantity ingested by humans. Along the same lines, it is worth mentioning the research conducted by Cohen et al. (1967), which set off the whole "chromosome breaks" scare: These scientists poured high concentrations of LSD on cultured cells and went on to show that the chromosomes of these cells featured twice as many breaks as normal. It has since been shown that substances in common use, such as milk, caffeine, and aspirin, lead to similar results at sufficient concentrations (see, for instance, Kato and Jarvik 1969). Dishotsky et al. (1971), who reviewed a total of 68 studies on the supposed effects of LSD on chromosomes, wrote in the conclusion of their article for *Science:* "From our own work and from a review of the literature, we believe that pure LSD ingested in moderate doses does not damage chromosomes in vivo, does not cause detectable genetic damage, and is not a teratogen or a carcinogen in man. Within these bounds, therefore, we suggest that, other than during pregnancy, there is no present contraindication to the continued controlled experimental use of pure LSD" (p. 439). Finally, see Yielding and Sterglanz (1968), Smythies and Antun (1969), and Wagner (1969) concerning the intercalation of LSD into DNA.

18. Yielding and Sterglanz (1968) write: "A study of the interactions between LSD and such macromolecules as DNA may also be relevant to the psychotomimetic actions of such drugs. . . . Thus, binding to DNA would appear to be a general property of this group of drugs" (p. 1096). This idea was taken further by McKenna and McKenna (1975) in a visionary speculation: "We speculated that information stored in the neural-genetic material might be made available to consciousness through a modulated ESR [electron spin resonance] absorption phenomenon, originating in superconducting charge-transfer complexes formed by intercalation of tryptamines and beta-carbolines into the genetic material. We reasoned that both neural DNA and neural RNA were involved in this process: Serotonin or, in the case of our experiment, exogenously introduced methylated tryptamines would preferentially bind to membrane RNA, opening

the ionic shutter mechanism and, simultaneously, entering into su-
perconductive charge transfer with its resulting modulated ESR signal; beta-carbolines could then pass through the membrane via the
RNA-ionic channel and intercalate into the neural DNA" (p. 104).
Dennis McKenna has since become an experienced researcher on
neurological receptors, but his work does not deal any further with
DNA. Terence McKenna (1993) tells the story behind the conception of these visionary speculations.

19. The advances accomplished over the last twenty-five years regarding science's understanding of neurological receptors can be gauged
by reading Smythies (1970) on the possible nature of these receptors: "This makes deductions from the chemical relation between
various agonists and antagonists to the possible nature of the receptor site tentative at best. Such arguments would be more cogent if
anything were known, on independent grounds, of the chemical nature of the receptor site. Unfortunately very little is known" (p. 182).
In those days, scientists could only advance on this question by grop-
ing in the dark; Symthies theorized, incorrectly, that the receptors
were made of RNA.

20. For instance, in the most recent edition of the *Psychedelics encyclopedia* (Stafford 1992), there is no reference to DNA. To my knowledge, the only other mention of a link between hallucinogens and
DNA is by Lamb (1985), who suggests in passing: "Perhaps on some
unknown unconscious level the genetic encoder DNA provides a
bridge to biological memories of all living things, an aura of un-
bounded awareness manifesting itself in the activated mind" (p. 2).
Lamb elaborates no further on this.

21. See Rattemeyer et al. (1981), Popp (1986), Li (1992), Van Wijk and
Van Aken (1992), Niggli (1992), Mei (1992), and Popp, Gu, and Li
(1994).

22. Popp (1986, p. 207).

23. Popp (1986, pp. 209, 207). See also Popp, Gu, and Li (1994) regarding the coherence in biophoton emission.

24. Suren Erkman, personal communication, 1995.

25. Strassman et al. (1994, pp. 100–101).

26. Etymologically, "hallucination" comes from the Latin *hallucinari,*
"to wander in the mind," which corresponds quite precisely to the
description I propose of the phenomenon induced by hallucinogens—namely, a shifting of consciousness away from ordinary real-

ity toward the molecular level. The word *hallucinari* only acquired the pejorative meaning "to be mistaken" in the fifteenth century; but I do not consider this connotation a sufficient reason not to use a word which is commonly understood and the original etymology of which corresponds to the described phenomenon. Finally, and in opposition to a certain number of current scholars, I do not subscribe to the use of the newly coined word "entheogen" (to replace "hallucinogen"), because it jargonizes a difficult subject and loads it with divine (*theos* = "God") connotations.

27. Popp, Gu, and Li (1994) write; "There is evidence of nonsubstantial biocommunication between cells and organisms by means of photon emission" (p. 1287). On biophoton emission as a cellular language, see Galle et al. (1991), Gu (1992), and Ho and Popp (1993). One of the most eloquent experiments in this field consists of placing two lots of unicellular organisms in a device which measures photon emission and separating them with a metal screen; under these circumstances, the graph of the first lot's photon emission shows no relationship to that of the second lot. When the metal screen is removed, both graphs coincide to the highest degree—see Popp (1992a, p. 40). On the role of biophoton emission in plankton colonies, see Galle et al. (1991).

28. Ho and Popp (1993, p. 192).

29. Fritz-Albert Popp, personal communication, 1995.

30. On the precursory work of Alexander A. Gurvich, see the references in Popp, Gu, and Li (1994) as well as the writings of Anna A. Gurvich (1992, for example).

31. Reichel-Dolmatoff (1979, p. 117). On the importance of quartz crystals in shamanic practices, see also Harner (1980, pp. 138–140) and Eliade (1972).

32. Baer (1992) writes concerning the use of quartz crystals by Matsigenka shamans: "Light-colored or transparent stones, especially quartz crystals, are regarded as curative. They are called *isere'pito*. Although this designation is the same as that for the auxiliary spirits, it is more correct to view them as 'bodies,' 'residences,' or material manifestations of these spirits. . . . The Matsigenka say the shaman feeds his stones tobacco daily. If he does not do so, his auxiliary spirits, which materialize in the crystals, will leave him, and then the shaman will die" (pp. 86–87). The same practice is found among neighboring Ashaninca sheripiári (see Elick 1969, pp. 208–209).

33. Frank-Kamenetskii (1993, p. 31).

34. Blocker and Salem (1994) write: "In DNA, one finds four bases which are different and all quite complex. The structure of two of these bases, thymine (T) and cytosine (C), is hexagonal. The other two, adenine (A) and guanine (G), have a nine atom structure, with a hexagon placed next to a pentagon" (p. 55).

35. While I suggest the hypothesis that DNA's "non-coding" repeat sequences serve, among other things, to pick up photons at different frequencies, it is worth mentioning that Rattemeyer et al. (1981) proposed, in the first article published on DNA as a source of photon emission, that the non-coding parts of the genome could play an unsuspected electromagnetic role: "Only a very small proportion (about 0.1 and 2%) of DNA operates as genetic material and is organized in nucleotide sequences according to the genetic code. Models have, therefore, been proposed which suggest some regulatory role for the non-protein-coding DNA. Recently, this regulatory role is being seen more in terms of some basic physical mechanisms, particularly the coherent electromagnetic interactions between different DNA sections, rather than a biochemical store of information" (p. 573). Li (1992, p. 190) also suggests that the aperiodic nature of the DNA crystal facilitates the coherence of photon emission. I suggest here that the converse is also true and that the repeat sequences in the DNA crystal facilitate its capacity to pick up photons.

36. Of course, biophoton researchers are aware of the fact that photon emission, considered as a cellular language, necessarily implies a receptor. Ho and Popp (1993) write that this phenomenon "points to the existence of amplifying mechanisms in the organisms receiving the information (and acting on it). Specifically, the living system itself must also be organized by intrinsic electrodynamical fields, capable of receiving, amplifying, and possibly transmitting electromagnetic information in a wide range of frequencies—rather like an extraordinarily efficient and sensitive, and extremely broadband radio receiver and transmitter, much as Fröhlich has suggested" (p. 194). I write that biophoton reception has not been studied, but Li (1992, p. 167) and Niggli (1992, p. 236) both mention in passing the necessary existence of a photon-trapping mechanism.

37. Chwirot (1992) writes: "The properties of chromatin [the substance contained in the nucleus—that is, DNA and its coating of proteins], optical ones included, are very different *in vivo* and *in vitro* and depend on many factors which have not yet been fully understood" (pp. 274–275). Popp, Gu, and Li (1994) conclude their review of the

biophoton literature by writing that "the mechanism [of biophoton emission] is not known in detail at present" (p. 1293).

38. Popp (1992b) writes: "The entity of all living systems (which can be considered as a more or less fully interlinked unit), rather than the individuals, is always developing" (p. 454).

10: BIOLOGY'S BLIND SPOT

1. Crick (1981, p. 58). Jones (1993) writes: "The ancestral message from the dawn of life has grown to an instruction manual containing three thousand million letters coded into DNA. Everyone has a unique edition of the manual which differs in millions of ways from that of their fellows. All this diversity comes from accumulated errors in copying the inherited message" (p. 79). Delsemme (1994) writes: "The mechanism [of evolution] is extraordinarily simple, as it rests on two principles: copying errors, which cause 'mutations'; survival of the individual best adapted to its environment" (p. 185). Francis Crick coined the term "central dogma" in 1958. Blocker and Salem (1994) write regarding the central dogma: "However, . . . this principle can be seriously challenged. In fact, from a certain point of view, one can almost consider it to be wrong: information actually flows back from the proteins to the genes, but by a different means, that of regulation" (p. 66). Regarding resistance to the theory of natural selection until the middle of the twentieth century, Mayr (1982) writes: "Up to the 1920s and 1930s, virtually all the major books on evolution—those of Berg, Bertalanffy, Beurlen, Böker, Goldschmidt, Robson, Robson and Richards, Schindewolf, Willis, and those of all the French evolutionists, including Cuénot, Caullery, Vandel, Guyénot, and Rostand—were more or less strongly anti-Darwinian. Among nonbiologists Darwinism was even less popular. The philosophers, in particular, were almost unanimously opposed to it, and this opposition lasted until relatively recent years (Cassirer, 1950; Grene, 1959; Popper, 1972). Most historians likewise rejected selectionism (Radl, Nordenskiöld, Barzun, Himmelfarb)" (p. 549). Mayr goes on to describe an international symposium held in 1947: "All participants endorsed the gradualness of evolution, the preeminent importance of natural selection, and the populational aspect of the origin of diversity. Not all other biologists were completely converted. This is evident from the great efforts made by Fisher, Haldane, and Muller as late as the 1940s and 50s to present again and

again evidence in favor of the universality of natural selection, and from some reasonably agnostic statements on evolution made by a few leading biologists such as Max Hartmann" (p. 569).

2. Crick (1966, p. 10) and Jacob (1974, p. 320).

3. Monod (1971, pp. 30–31).

4. Jakobson (1973, p. 61). He also writes: "Consequently, we can say that, of all the information-transmitting systems, the genetic code and the verbal code are the only ones that are founded on the use of discrete elements, which are, in themselves, devoid of meaning, but which are used to constitute the minimal units of significance, namely the entities endowed with a meaning that is their own in the code in question" (p. 52). See Shanon (1978) on the *differences* between the genetic code and human languages.

5. Calladine and Drew (1992) write: "The mass of DNA is surrounded in most cells by a strong membrane with tiny, selective holes, that allow some things to go in and out, but keep others either inside or outside. Important chemical molecules go in and out of these holes, like memos from the main office of a factory to its workshops; and indeed the individual cell is in many ways like an entire factory, on a very tiny scale. The space in the cell which is not occupied by DNA and the various sorts of machinery is filled with water" (p. 3). De Rosnay (1966) writes: "The cell is, indeed, a veritable molecular factory, but this 'miracle' factory is capable not only of looking after its own maintenance—as we have just seen—but also of building its own machines as well as the drivers of those machines" (p. 62). Pollack (1994) compares a cell to a city, rather than to a factory: "A cell is a busy place, a city of large and small molecules all constructed according to information encoded in DNA. The metaphor of a city may seem even more farfetched than that of a skyscraper for an invisibly small cell until you consider that a cell has room for more than a hundred million million atoms; that is plenty of space for millions of different molecules, since even the largest molecules in a cell are made of only a few hundred million atoms" (p. 18). In his book *The machinery of life*, Goodsell (1993) writes: "Like the machines of our modern world, these molecules are built to perform specific functions efficiently, accurately, and consistently. Modern cells build hundreds of thousands of different molecular machines, each performing one of hundreds of thousands of individual tasks in the process of living. These molecular machines are built according to four basic molecular plans. Whereas our macrosocopic machines

are built of metal, wood, plastic and ceramic, the microscopic machines in cells are built of protein, nucleic acid, lipid, and polysaccharide. Each plan has a unique chemical personality ideally suited to a different role in the cell" (p. 13). De Rosnay (1966, p. 165) compares enzymes to "biological micro-computers" and to "molecular robots," whereas Goodsell (1993, p. 29) calls them "automata." Wills (1991) writes: "The genome is like a book that contains, among many other things, detailed instructions on how to build a machine that can make copies of it—and also instructions on how to build the tools needed to make the machine" (p. 41). For discussions of DNA as a "language" or a "text," see, for example, Frank-Kamenetskii (1993, pp. 63–74), Jones (1993), or Pollack (1994). Atlan and Koppel (1990) reject the classical metaphor of DNA as a "program" and suggest instead that it is better understood as "data to a program embedded in the global geometrical and biochemical structure of the cell" (p. 338). Finally, Delsemme (1994, p. 205) writes that "we can consider with complete peace of mind that life is a normal physicochemical phenomenon."

6. Piaget (1975) writes: "Thus the most developed science remains a continual becoming, and in every field nonbalance plays a functional role of prime importance since it necessitates re-equilibration" (p. 178).

7. Scott quoted in Freedman (1994), whose article inspired this paragraph. Goodsell (1993) writes that "proteins are self-assembling machines," which, among other functions, "form motors, turning huge molecular oars that propel bacterial cells" or "specific pumps [that] are built to pump amino acids in, to pump urea out, or to trade sodium for potassium" (pp. 18, 42).

8. Calladine and Drew (1992, p. 37). See Wills (1989, p. 166) on the speed of carbonic anhydrase. See Radman and Wagner (1988, p. 25) on the minute rate of error of repair enzymes. *Science* nominated DNA repair enzymes "molecules of the year 1994." Recently, it was found that these enzymes are highly adaptable and that "repair" enzymes also participate in DNA replication, the control of the cell cycle, and the expression of genes. Similarly, enzymes that splice the double helix can do so in both chromosome recombination and repair operations. Enzymes that unwind DNA can act during transcription of the genetic text as well as repair (see Culotta and Koshland 1994). Wills (1991) writes on the speed of DNA duplication by enzymes called replisomes: "Replisomes work in pairs. As we

watch, about 100 pairs of replisomes seize specific places on each of the chromosomes, and each pair begins to work in opposite directions. Since all the chromosomes are being duplicated at once, there are about ten thousand replisomes operating throughout the nucleus. They work at incredible speed, spewing out new DNA strands at the rate of 150 nucleotides per second. . . . At full bore, the DNA can be replicated at one and a half million nucleotides per second. Even at this rate, it would still take about half an hour to duplicate all six billion nucleotides" (pp. 113–114).

9. Margulis and Sagan (1986, p. 145). Since the time of writing the French original of this book, two articles by Heald et al. (1996) and Zhang and Nicklas (1996) seem to indicate that the dance of chromosomes is orchestrated by spindle microtubules, which function even in the absence of chromosomes. This does not remove the question of intention, however. As Hyams (1996, p. 397) comments: "A great many questions about mitosis remain to be answered. To what extent do chromosomes contribute to spindle formation and to their own movement at anaphase? Do they have a role in positioning the cleavage furrow? What holds sister chromatids together, how are they 'unglued.' and what is the signal for this detachment? How do the checkpoints that sense a single detached chromosome or an imperfect one work?"

10. Wade (1995a) writes: "Only DNA endures. This thoroughly depressing view values only survival, which the DNA is not in a position to appreciate anyway, being just a chemical" (p. 20).

11. Trémolières (1994, p. 138) considers that "our human comprehension and intelligence reach their own limits. It seems that our brain is one of the most complex objects that we can find in the universe." McGinn (1994, p. 67) writes: "We want to know, among other things, how our consciousness levers itself out of the body. We want, that is, to solve the mind-body problem, the deep metaphysical question about how mind and matter meet. But what if there is something about us that makes it impossible for us to solve this ancient conundrum? What if our cognitive structure lacks the resources to provide the requisite theory?"

12. Hunt (1996) writes: "Crow tool manufacture had three features new to tool use in free-living nonhumans: a high degree of standardization, distinctly discrete tool types with definite imposition of form in tool shaping, and the use of hooks. These features only first appeared in the stone and bone tool-using cultures of early humans

after the Lower Paleolithic, which indicates that crows have achieved a considerable technical capability in their tool manufacture and use" (p. 249). See Huffman (1995) on chimpanzees using medicinal plants. Perry (1983) writes about ants that herd aphids: "In one species, the ants take fine earth up to the leaves and stems of plants and, using their own saliva, cement together tiny shelters, shaped like mud huts, for their aphid partners. These shelters help to protect the aphids from severe weather and to some extent from predators. . . . Some ants will round up local populations of aphids at the end of the day, in much the same way that a sheepdog herds sheep. The ants then take their aphids down into the nest for protection from predators. In the morning the aphids are escorted to the required plant for another day's feeding and milking" (pp. 28–29). See also Hölldobler and Wilson (1990, pp. 522–529). Concerning mushroom-cultivating ants, see Chapela at al. (1994) and Hinkle et al. (1994). Wilson (1984, p. 17) compares an ant's brain to a grain of sugar.

13. Monod (1971, p. 18). Wesson (1991) writes: "By what devices the genes direct the formation of patterns of neurons that constitute innate behavioral patterns is entirely enigmatic. Yet not only do animals respond appropriately to manifold needs; they often do so in ways that would seem to require something like forethought" (p. 68). He adds: "An instinct of any complexity, linking a sequence of perceptions and actions, must involve a very large number of connections within the brain or principal ganglia of the animal. If it is comparable to a computer program, it must have the equivalent of thousands of lines. In such a program, not merely would chance of improvement by accidental change be tiny at best. It is problematic how the program can be maintained without degradation over a long period despite the occurrence from time to time of errors by replication" (p. 81). On the absence of a goal, or teleology, in nature, Stocco (1994) writes that "biological evolution does not proceed in a precise direction and aims at no particular goal" (p. 185), and Mayr (1983) writes: "The one thing about which modern authors are unanimous is that adaptation is not teleological, but refers to something produced in the past by natural selection" (p. 324). According to Wesson (1991): "For a biologist to call another a teleologist is an insult" (p. 10).

14. According to several recent studies, non-coding DNA might actually play a structural role and display the characteristics of a language,

the meaning of which remains to be determined. See Flam (1994), Pennisi (1994), Nowak (1994), and Moore (1996).

15. The twenty amino acids used by nature to build proteins vary in shape and function. Some play structural roles, such as making a hairpin turn that folds the protein back on itself. Others make sheet-like surfaces as docking sites for other molecules. Others form links between protein chains. Three amino acids contain benzene, a greasy compound that is the molecular equivalent of Velcro and that can hold certain substances and then release them without modifying its own structure. One finds these benzene-containing amino acids at exactly the right place in the "lock" of nicotinic receptors, where they bond molecules of acetylcholine or nicotine (see Smith 1994). Couturier et al. (1990) provide the exact sequence of the 479 amino acids that constitute one of the five protein chains of the nicotinic receptor. My estimate of 2,500 amino acids for the entire receptor is an extrapolation based on their work. See Lewis et al. (1987) regarding the presence of nicotinic receptors among nematodes.

16. Wesson (1991, p. 15).

17. Trémolières (1994, p. 51). He adds: "We know that more than 90% of the changes affecting a letter in a word of the genetic message lead to disastrous results; proteins are no longer synthesized correctly, the message loses its entire meaning and this leads purely and simply to the cell's death. Given that mutations are so frequently highly unfavourable, and even deadly, how can beneficial evolution be attained?" (p. 43). Likewise, Frank-Kamenetskii (1993) writes: "It is clear, therefore, that you need a drastic refitting of the whole of your machine to make the car into a plane. The same is true for a protein. In trying to turn one enzyme into another, point mutations alone would not do the trick. What you need is a substantial change in the amino acid sequence. In this situation, rather than being helpful, selection is a major hindrance. One could think, for instance, that by consistently changing amino acids one by one, it will eventually prove possible to change the entire sequence substantially and thus the enzyme's spatial structure. These minor changes, however, are bound to result eventually in a situation in which the enzyme has ceased to perform its previous function but it has not yet begun its 'new duties.' It is at this point that it will be destroyed—together with the organism carrying it" (p. 76).

18. Nash (1995, 68, 70).

19. See Wesson (1991, p. 52). He adds: "By Mayr's calculation, in a rapidly evolving line an organ may enlarge about 1 to 10 percent per million years, but organs of the whale-in-becoming must have grown ten times more rapidly over 10 million years. Perhaps 300 generations are required for a gene substitution. Moreover, mutations need to occur many times, even with considerable advantage, in order to have a good chance of becoming fixed. Considering the length of whale generations, the rarity with which the needed mutations are likely to appear, and the multitude of mutations needed to convert a land mammal into a whale, it is easy to conclude that gradualist natural selection of random variations cannot account for this animal" (p. 52). Wesson's book is a catalogue of biological improbabilities—from bats' hypersophisticated echolocation system to the electric organs of fish—and of the gaping holes in the fossil record.

20. Mayr (1988, pp. 529–530). Goodwin (1994) writes: "New types of organism appear upon the evolutionary scene, persist for various periods of it, and then become extinct. So Darwin's assumption that the tree of life is a consequence of the gradual accumulation of small hereditary differences appears to be without significant support. Some other process is responsible for the emergent properties of life, those distinctive features that separate one group of organisms from another, such as fishes and amphibians, worms and insects, horsetails and grasses. Clearly something is missing from biology" (p. x).

21. Shapiro (1996, p. 64).

22. *Mycoplasma genitalium* is the smallest genome currently known, at 580,000 base pairs. Mushegian and Koonin (1996) compared it to the genome of bacterium *Hemophilus influenzae*, which contains 1,800,000 base pairs, and concluded that the minimal amount of genetic information necessary for life is 315,000 base paris. This is still an enormous amount of information.

23. See Butler (1996) on the 12 million base pair genome of the yeast *Saccharomyces cerevisiae*. See Hilts (1996) on the similarities between yeast and human genes. In some cases, the contrary is also true, and genomes vary greatly between closely related species; Wade (1997b) writes about a conference on small genomes: "As work on one genome after another was described at the meeting, the scientists' mood was like that of people looking at newly-discovered treasure maps, with the treasure not yet in hand but with wonderfully tantalizing clues all about. For example, the order of genes in a

genome seems to vary widely, even between closely related species of microbes, as if evolution were constantly shuffling the deck" (p. A14).

24. Langaney (1997, p. 122). Holder and McMahon (1996) write: "Remarkably, many of the genes that are important for the control of fly development are also crucial players in vertebrate, and by association human, development. . . . Some of the similarities are amazing: for example, mutations in both human *Pax6* gene and in *eyeless,* the *Drosophila* homologue, cause abnormal eye development. This maintenance of function occurs in spite of the overtly different manner in which *Drosophila* and human eyes develop" (p. 515). Yoon (1995) writes: "From silken-petaled roses to popping snapdragons to a willow tree's fuzzy catkins, the plant world offers a dazzling array of flowers. Yet the difference between all this blooming beauty and a plain green shoot appears to be nothing more than the flicking on of one master genetic switch, according to two new studies. Using genetically engineered plants, researchers were able to show that either of two genes, on its own, could turn on the cascade of thousands of genes that produce a flower. Researchers were able to use the genes . . . to produce blossoms where there should instead have been leafy shoots in plants as diverse as Arabidopsis, a roadside weed, tobacco and aspen trees" (p. B5). Wade (1997c) writes: "Many of the most important fruit fly genes, like those that tell the developing embryo to produce organs at certain places, have been found to have counterparts in humans. The fly and human versions of these genes are not identical but have recognizably similar DNA sequences, reflecting their descent from a common ancestral gene some 550 million years ago"; he also writes that there is "surprising and extensive overlap of the genes among all the model organisms" (p. B7). Biology's main model organisms are fruit fly, mouse, worm *C. elegans,* zebra fish, and human.

25. See Hilts (1996, p. C19) on genes "that appear to clump together in families that work on similar problems." See Wade (1997a) on the similarities in gene clusters on mouse and human X chromosomes.

26. Pollack (1997, p. 674).

27. Luisi (1993, p. 19) and Popper (1974, pp. 168, 171). Popper (1974) writes: "I now wish to give some reasons why I regard Darwinism as metaphysical, and as a research programme. It is metaphysical because it is not testable. One might think that it is. It seems to assert that, if ever on some planet we find life which satisfies conditions (a)

and (b) [heredity and variation], then (c) [natural selection] will come into play and bring about in time a rich variety of distinct forms. Darwinism, however, does not assert as much as this. For assume we find life on Mars consisting of exactly three species of bacteria with a genetic outfit similar to that of three terrestrial species. Is Darwinism refuted? By no means. We shall say that these three species were the only forms among the many mutants which were sufficiently well adjusted to survive. And we shall say the same if there is only one species (or none). Thus Darwinism does not really *predict* the evolution of variety. It therefore cannot really *explain* it. At best, it can predict the evolution of variety under 'favourable conditions.' But it is hardly possible to describe in general terms what favourable conditions are—except that, in their presence, a variety of forms will emerge" (p. 171, original italics). Dawkins (1986) provides a good illustration of the tautologous tendencies of Darwinism when he writes: "Even if there were no actual evidence in favour of the Darwinian theory (there is, of course) we should still be justified in preferring it over all rival theories" (p. 287). He also tells a charming story of a beaver that undergoes a point mutation in its genetic text; this leads to a change in the beaver's brain's "wiring diagram," which makes the beaver hold its head higher in the water while swimming with a log in its mouth; this makes it less likely that the mud washes off the log, which makes the log stickier, which makes the beaver's dam a sounder structure, which increases the size of the lake, which makes the beaver's lodge more secure against predators, which increases the number of offspring reared by the beavers. This means that beavers with the mutated gene will become more numerous in time and will eventually become the norm. He concludes: "The fact that this particular story is hypothetical, and that the details may be wrong, is irrelevant. The beaver dam evolved by natural selection, and therefore what happened cannot be very different, except in practical details, from the story I have told" (p. 136). Wilson (1992) even provides an explicitly Darwinian explanation for the worldwide phenomenon of snake veneration, thereby showing that the theory of natural selection can be used to justify more or less anything: "People are both repelled and fascinated by snakes, even when they have never seen one in nature. In most cultures the serpent is the dominant wild animal of mythical and religious symbolism. Manhattanites dream of them with the same frequency as Zulus. This response appears to be Darwinian in origin. Poisonous snakes have been an important cause of mortality almost everywhere, from Finland to Tasmania, Canada to Patagonia; an untutored alertness in their presence saves lives. We note a kindred response in many primates, including Old World monkeys and chimpanzees" (p. 335). See also Moorhead and Kaplan, eds. (1967), Chandebois (1993), and Schützenberger (1996) on the limits of Darwinism.

11: "WHAT TOOK YOU SO LONG?"

1. Jacques Mabit, a medical doctor doing remarkable work with mestizo ayahuasqueros in Peru, notes that in the ayahuasca literature, which contains over five hundred titles, less than 10 percent of the authors have tried the substance, and none has followed the classical apprenticeship (see Mabit et al. 1992). Mabit himself is one of the rare exceptions.

2. Hill (1992), in his article on Wakuénai musical curing, writes regarding the fragmentation of Western knowledge: "Wakuénai curing rituals are simultaneously musical, cosmological, social, psychological, medical, and economic events, a multidimensionality that 'embarrasses the categories' of Western scientific and artistic culture" (p. 208).

3. Regarding the failure of Western-style education among the indigenous people of Amazonia, see Gasché (1989–1990). Moreover, Gasché points out that intercultural education requires not only funds, but a calling into question of anthropology as a science, given that the discipline bases its existence on intercultural dialogue between Indians and non-Indians, which can only occur through a constant confrontation of these two realities; up until now, an anthropology that is truly useful to the people who are its object remains to be realized. Thus, Gasché (1993) writes: "From a strictly logical, or more precisely topological, point of view, one can envisage the orientation of anthropological discourse in the direction not of the researcher's own society, but, on the contrary, of the society which is, or was, its object of study. Such a proposition no doubt surprises, or even shocks some anthropologists, because, indeed, it has hardly been formulated and has even less led to careers. However, for anthropologists who assume the principle of cultural relativism as a presupposition founding their scientific attitude towards human societies, this proposition would logically emerge as soon as they

postulate the coherence between their scientific statements and their social actions: if all societies are of equal worth, why do anthropologists keep the benefits of the product of their labor exclusively for their own society? This question is all the more urgent that it brings into play two other central notions in anthropology, namely exchange and reciprocity: the data, which are the raw material of all anthropological thought, come from the society that never benefits from the finished product. And it is the question of return, of equilibration in the relationship between the Indian society and the anthropologist, between the object and subject of the research, which many Indians are currently posing in the Peruvian Amazon" (pp. 27–28).

4. Davis (1993) writes: "The current international discussion of biodiversity prospecting and intellectual property rights fails to comprehend this sacred or spiritual quality of Indigenous plant knowledge, because it is so rooted in material considerations and the economic thinking of the West" (p. 21). Posey (1994) writes: "Intellectual property rights is a foreign concept to indigenous peoples" (p. 235).

5. Luna and Amaringo (1991, p. 72). Regarding the multicultural past of Pablo Amaringo, see p. 21 of the same book.

6. See Taussig (1987, p. 179).

7. Chaumeil (1992) writes: "We know about the fascination that the forest and its inhabitants exert in matters of shamanism on Andean and urban society. Urban and Andean shamans generally attribute great powers to their indigenous colleagues, whom they visit frequently, setting up vast shamanic exchange networks in Colombia, Ecuador and Peru. In Brazil, many *mestizo* shamans adopt indigenous methods and live temporarily in Indian villages to learn the shamanic arts. Indeed, most claim to have had at least one indigenous instructor, or recognize the indigenous origin of their knowledge" (p. 93). Chaumeil goes on to explain that this exchange works both ways and that there is "an increasing flux of young indigenous people into towns where they go to learn the shamanic arts with *mestizo* instructors, who develop the opposite tendency" (p. 99).

8. Rosaldo (1980) writes: "Doing oral history involves telling stories about stories people tell about themselves. Method in this discipline should therefore attend to 'our' stories, 'their' stories, and the connections between them" (p. 89). Rosaldo (1989) writes: "Such terms as *objecivity, neutrality,* and *impartiality* refer to subject positions once endowed with great institutional authority, but they are

arguably neither more nor less valid than those of more engaged, yet equally perceptive, knowledgeable social actors" (p. 21, original italics). He adds: "Because researchers are necessarily both somewhat impartial and somewhat partisan, somewhat innocent and somewhat complicit, their readers should be as informed as possible about what the observer was in a position to know and not know" (p. 69).

9. "Learned analysis" often escapes the understanding not only of those who are its object, but of many Western individuals. Anthropologists have written so many unreadable texts that the literary critic Pratt (1986) writes: "For the lay person, such as myself, the main evidence of a problem is the simple fact that ethnographic writing tends to be surprisingly boring. How, one asks constantly, could such interesting people doing such interesting things produce such dull books? What did they have to do to themselves?" (p. 33).

10. For a detailed discussion of the role of intuition, dreaming, imagination, and illumination in the history of scientific discoveries, see Beveridge (1950). Watson (1968) writes: "Afterwards, in the cold, almost unheated train compartment, I sketched on the blank edge of my newspaper what I remembered of the B pattern. Then as the train jerked towards Cambridge, I tried to decide between two- and three-chain models. As far as I could tell, the reason the King's group did not like two chains was not foolproof. It depended upon the water content of the DNA samples, a value they admitted might be in great error. Thus by the time I had cycled back to college and climbed over the back gate, I had decided to build two-chain models. Francis would have to agree. Even though he was a physicist, he knew that important biological objects come in pairs" (p. 166). The "B structure" mentioned by Watson refers to an X-ray photograph of DNA taken by Rosalind Franklin, whose work was thus central to Watson and Crick's discovery, but who received no mention when the Nobel Prize was awarded. That she was a woman, and that things should have occurred this way, was surely no coincidence.

11. Beveridge (1950, p. 72). He adds: "The most important prerequisite is prolonged contemplation of the problem and the data until the mind is saturated with it. There must be a great interest in it and desire for its solution. The mind must work consciously on the problem for days in order to get the subconscious mind working on it. . . . An important condition is freedom from other problems or interests competing for attention, especially worry over private affairs. . . . Another favourable condition is freedom from interruption or even

fear of interruption or any diverting influence such as interesting conversation within earshot or sudden and excessively loud noises. . . . Most people find intuitions are more likely to come during a period of apparent idleness and temporary abandonment of the problem following periods of intensive work. Light occupations requiring no mental effort, such as walking in the country, bathing, shaving, travelling to and from work, are said by some to be when intuitions most often appear. . . . Others find lying in bed most favourable and some people deliberately go over the problem before going to sleep and others before rising in the morning. Some find that music has a helpful influence but it is notable that only very few consider that they get any assistance from tobacco, coffee or alcohol" (p. 76). Mullis (1994) discusses in his Nobel lecture how he conceived the polymerase chain reaction while driving along a moonlit mountain road with his driving companion asleep next to him. The polymerase chain reaction allows one to amplify DNA from a few cells to vat fulls of cells in a few hours; it spawned the genetic engineering revolution.

12. Artaud (1979, p. 193). The French original is "Je me livre à la fièvre des rêves, mais c'est pour en retirer de nouvelles lois."

13. The contents of this famous soup are problematic. In 1952, Stanley Miller and Harold Urey did an experiment that was to become famous; they bombarded a test tube containing water, hydrogen, ammonia, and methane with electricity, supposedly imitating the atmosphere of the primitive earth with its permanent lightning storms; after a week, they had produced 2 of the 20 amino acids that nature uses in the construction of proteins. This experiment was long cited as proof that life could emerge from an inorganic soup. However, in the 1980s, geologists realized that an atmosphere of methane and ammoniac would rapidly have been destroyed by sunlight and that our planet's primitive atmosphere most probably contained nitrogen, carbon dioxide, water vapor, and traces of hydrogen. When one bombards the latter with electricity, one does not obtain biomolecules. So the prebiotic soup is increasingly considered to be a "myth" (see Shapiro 1986).

14. Reisse (1988) writes about panspermia "that this theory presents a major defect. No acceptable criterion allows one to measure its quality: by essence it cannot be refuted. Moreover, panspermia in its modern version displaces the location where life originated but leaves the fundamental problem of its origin intact" (p. 101). De

Duve (1984) writes: "If you equate the probability of the birth of a bacterial cell to that of the chance assembly of its component atoms, even eternity will not suffice to produce one for you. So you might as well accept, as do most scientists, that the process was completed in no more than 1 billion years and that it took place entirely on the surface of our planet, to produce, as early as 3.3 billion years ago, the bacteriumlike organisms revealed by fossil traces" (p. 356). Watson et al. (1987) write in their chapter on the origins of life: "In this chapter, we will assume, as do the vast majority of practicing biologists, that life originated on Earth" (p. 1098).

15. In the early 1980s, researchers discovered that certain RNA molecules, called "ribozymes," could cut themselves up and stick themselves back together again, acting as their own catalysts. This led to the following speculation: If RNA is also an enzyme, it could perhaps replicate itself without the help of proteins. An RNA that is both gene and catalyst would solve the old chicken-and-egg problem that has haunted the debate on the origin of DNA and proteins. Scientists went on to formulate the theory of the "RNA world," according to which the first organisms were RNA molecules that learned to synthesize proteins, facilitating their replication, and that surrounded themselves with lipids to form a cellular membrane; these RNA-based organisms then evolved into organisms with a genetic memory made of DNA, which is more stable chemically. However, this theory is not only irrefutable, it leaves many questions unsolved. Thus, to make RNA, one must have nucleotides, and for the moment, no one has ever seen nucleotides take shape by chance and line up to form RNA. As Shapiro (1994b) writes, the "experiments conducted up until now have shown no tendency for a plausible prebiotic soup to build bricks of RNA. One would have liked to discover ribozymes capable of doing so, but this has not been the case. And even if one were to discover any, this would still not resolve the fundamental question: where did the first RNA molecule come from?" (pp. 421–422). He adds: "After ten years of relentless research, the most common and remarkable property of ribozymes has been found to be the capacity to demolish other molecules of nucleic acid. It is difficult to imagine a less adapted activity than that in a prebiotic soup where the first colony of RNA would have had to struggle to make their home" (p. 421). Kauffman (1996) writes: "The dominant view of life assumes that self-replication must be based on something akin to Watson-Crick base pairing. The 'RNA

world' model of the origins of life conforms to this view. But years of careful effort to find an enzyme-free polynucleotide system able to undergo replication cycles by sequentially and correctly adding the proper nucleotide to the newly synthesized strand have not yet succeeded" (p. 497). Laszlo (1997) writes: "The origin of life is more a question of metaphysics than a scientific problem. The experimental facts gleaned by different well-established authors allow only for scenarios, in an unlimited number, all of which are fictive" (p. 26). Regarding clay-based speculations, see Cairns-Smith (1983); regarding oily bubbles, see Morowitz (1985); regarding self-replicating peptides, see Lee et al. (1996).

16. Trémolières (1994) writes: "Despite these terrible paradoxes, the scientific world agrees that there must have existed something before the current organization of life, and more precisely that there were 'living' or 'pre-living' forms that did not yet contain the genetic code, or in any case, not the code that we know. And science has strangely developed its branches in a direction where nothing exists any longer; this is the contrary of futurology—which is apparently a science—or of science fiction, which is an art" (p. 70). Shapiro (1986) writes: "Scientific explanations flounder, however, and possibilities multiply when we ask how this first cell arose on earth. Competing theories abound—which seems always the case when we know very little about a subject. Some theories, of course, come labeled as The Answer. As such they are more properly classified as mythology or religion than as science" (p. 13).

17. Shapiro (1994a, p. II). Watson et al. (1987) write: "Unfortunately, it is impossible to obtain direct proof for any particular theory of the origin of life. The sobering truth is that even if every expert in the field of molecular evolution were to agree on how life originated, the theory would still be a best guess rather than a fact" (p. 1161). Wade (1995c) writes: "With a handful of trivial exceptions, all forms of life have the same, apparently arbitrary code through which DNA specifies protein molecules. If life arises so spontaneously, why don't we see a variety of different codes and chemistries in earth's creatures? The universal nature of the genetic code implies a one-time event, some narrow gateway through which only a single entity or family of related life forms was able to pass. One possibility is that life evolved independently several times on earth and creatures with our genetic code destroyed those based on all other codes. But there's no evidence for such a code war. Or maybe the emergence of

life is indeed so improbable that it only happened once. Strange, then, that life seems to have arisen at the earliest moment possible, almost immediately after the primitive earth had cooled enough" (pp. 22–23).

18. Sullivan (1988, p. 33).
19. Chuang-Tzu (1968, p. 43).

BIBLIOGRAPHY

Abelin, Theodor. 1993. Tabac et santé. Berne: Association Suisse contre la Tuberculose et les maladies Pulmonaires (ASTP).

Alberts, Bruce, et al. 1990. Biologie moléculaire de la cellule. Paris: Flammarion Médecines-Science, 2d ed.

Artaud, Antonin. 1979. Manifeste en langage clair. In *L'ombilic des limbes*, pp. 192–194. Paris: Gallimard (original text, 1925).

Atkinson, Jane. 1992. Shamanisms today. *Annual Review of Anthropology*, 21:307–330.

Atlan, Henri, and Moshe Koppel. 1990. The cellular computer DNA: Program or data. *Bulletin of Mathematical Biology*, 52(3):335–348.

Baer, Gerhard. 1992. The one intoxicated by tobacco: Matsigenka shamanism. In *Portals of power: Shamanism in South America*, E. Jean Matteson Langdon and Gerhard Baer, eds., pp. 79–100. Albuquerque: University of New Mexico Press.

Balick, Michael J., Elaine Elisabetsky, and Sarah A. Laird, eds. 1996. *Medicinal resources of the tropical forest: Biodiversity and its importance to human health*. New York: Columbia University Press.

Barker, Steven A., et al. 1981. N,N-Dimethyltryptamine: An endogenous hallucinogen. *International Review of Neurobiology*, 22:83–110.

Bass, Thomas A. 1994. *Reinventing the future: Conversations with the world's leading scientists*. New York: Addison-Wesley.

Baudoin, Marcel. 1918. *La préhistoire du caducée*. Paris: Imprimerie de la Bourse du Commerce.

Bayard, Jean-Pierre. 1987. *Le symbolisme du caducée*. Paris: Trédaniel.

Beach, C., et al. 1994. Electron migration along 5-bromouracil-substituted DNA irradiated in solution and in cells. *Radiation Research*, 137(3):385–393.

Beauclerk, John, and Jeremy Narby, with Janet Townsend. 1988. Indigenous peoples: A fieldguide for development. Oxford: OXFAM.

Bellier, Irène. 1986. Los cantos Mai Huna del yagé. *América Indígena*, 46(1):129–148.

Beveridge, W.I.B. 1950. *The art of scientific investigation*. London: Heinemann.

Bisset, N. G. 1989. Arrow and dart poisons. *Journal of Ethnopharmacology*, 25:1–41.

Blocker, Ariel, and Lionel Salem. 1994. *L'homme génétique*. Paris: Dunod.

Blubaugh, L. V., and C. R. Linegar. 1948. Curare and modern medicine. *Economic Botany*, 2:73–82.

Bormann, F. H., and S. R. Kellert, eds. 1990. *Ecology, economics, ethics: The broken circle*. New Haven: Yale University Press.

Boulnois, Jean. 1939. *Le caducée et la symbolique dravidienne, de l'arbre, de la pierre et de la déesse-mère*. Paris: Librairie d'Amérique et d'Orient.

Bourdieu, Pierre. 1977. *Outline of a theory of practice*. New York: Cambridge University Press.

———. 1980. *Le sens pratique*. Paris: Editions de Minuit.

———. 1990. *The logic of practice*. Stanford, CA: Stanford University Press.

Bourguignon, Erika. 1970. Hallucination and trance: An anthropologist's perspective. In *Origin and mechanisms of hallucinations*, Wolfram Keup, ed., pp. 183–190. New York: Plenum Press.

Broad, William J. 1994. Microbial life deep in the planet. *International Herald Tribune*, 6 October, p. 10.

Browman, D. L., and R. A Schwarz, eds. 1979. *Spirits, shamans and stars: Perspectives from South America*. Paris–The Hague: Mouton Publishers.

Brown, Michael Forbes. 1988. Shamanism and its discontents. *Medical Anthropology Quarterly*, 2:102–120.

Buchillet, Dominique. 1982. Recension de "Michael Harner, *Chamane. Les secrets d'un sorcier indien d'Amérique du Nord.*" *L'Ethnographie*, 87–88(1–2):259–261.

Burnand, Christiane. 1991. *La coupe et le serpent*. Nancy: Presses Universitaires de Nancy.

Burroughs, William, and Allen Ginsberg. 1963. *The yagé letters*. San Francisco: City Lights.

Butler, Declan. 1996. Interest ferments in yeast genome sequence. *Nature*, 380:660–661.

Cairns-Smith, Graham. 1983. *Genetic takeover.* Cambridge: Cambridge University Press.

Calladine, Chris, and Horace R. Drew. 1992. *Understanding DNA.* London: Academic Press.

Campbell, Joseph. 1959. *The masks of God: Primitive mythology.* New York: Arkana.

———. 1964. *The masks of God: Occidental mythology.* New York: Arkana.

———. 1968. *The masks of God: Creative mythology.* New York: Arkana.

Chandebois, Rosine. 1993. *Pour en finir avec le darwinisme: Une nouvelle logique du vivant.* Montpellier: Editions Espaces 34.

Changeux, Jean-Pierre. 1983. *L'homme neuronal.* Paris: Fayard.

———. 1993. Chemical signaling in the brain. *Scientific American,* 268(4):30–37.

Chapela, Ignacio H., et al. 1994. Evolutionary history of the symbiosis between fungus-growing ants and their fungi. *Science,* 266:1691–1694.

Chaumeil, Jean-Pierre. 1982. Les plantes-qui-font-voir. Rôle et utilisation des hallucinogènes chez les Yagua du Nord-Est péruvien. *L'Ethnographie,* 87–88 (2–3):55–84.

———. 1983. *Voir, savoir, pouvoir. Le chamanisme chez les Yagua du Nord-Est péruvien.* Paris: Editions de l'Ecole des Hautes Etudes en Sciences Sociales.

———. 1992. Chamanismes à géométrie variable en Amazonie. *Diogène,* 158:92–103.

———. 1993. Des esprits aux ancêtres: Procédés linguistiques, conceptions du langage et de la société chez les Yagua de l'Amazonie péruvienne. *L'Homme,* 126–128(2–4):409–427.

Chevalier, Jacques M. 1982. *Civilization and the stolen gift: Capital, kin and cult in Eastern Peru.* Toronto: University of Toronto Press.

Chevalier, Jean, and Alain Gheerbrant. 1982. *Dictionnaire des symboles.* Paris: Robert Laffont.

Christensen, Jon, and Jeremy Narby. 1992. Indians back US on biodiversity. *Jornal do Brasil* (English ed.), June 12, p. 10.

Chuang-Tzu. 1968. *The complete works of Chuang-Tzu,* translated by Burton Watson. New York: Columbia University Press.

———. 1981. *The inner chapters,* translated by A. C. Graham. London: Mandala.

Chwirot, Barbara W. 1992. Ultraweak luminescence studies of microsporogenesis in larch. In *Recent advances in biophoton research and its applications,* F. A. Popp et al., eds., pp. 259–285. Singapore: World Scientific.

Cimino, M., et al. 1992. Distribution of nicotinic receptors in cynomolgus monkey brain and ganglia: Localization of α 3 subunit mRNA, α-bungarotoxin and nicotinic binding sites. *Neuroscience,* 51(1):77–86.

Clark, Robert Thomas Rundle. 1959. *Myth and symbol in Ancient Egypt.* London: Thames and Hudson.

Clery, Daniel. 1995. DNA goes electric. *Science,* 267:1270.

Clifford, James, and George E. Marcus, eds. 1986. *Writing culture: The poetics and politics of ethnography.* Berkeley: University of California Press.

Cohen, Maimon M., et al. 1967. Chromosomal damage in human leukocytes induced by lysergic acid diethylamide. *Science,* 155:1417–1419.

Colchester, Marcus. 1982. The cosmovision of the Venezuelan Sanema. *Antropológica,* 58:97–122.

Couturier, Sabine et al. 1990. A neuronal nicotinic acetylcholine receptor subunit (α 7) is developmentally regulated and forms a homo-oligomeric channel blocked by α-BTX. *Neuron,* 5:847–856.

Crick, Francis. 1958. On protein synthesis. *Symposium of the Society for Experimental Biology,* 12:548–555.

———. 1966. *Of molecules and men.* Seattle: University of Washington Press.

———. 1981. *Life itself: Its origin and nature.* London: MacDonald and Co.

———. 1994. The astonishing hypothesis: The scientific search for the soul. New York: Simon and Schuster.

Culotta, Elizabeth, and Daniel E. Koshland, Jr. 1994. Molecule of the year—DNA repair works its way to the top. *Science,* 266:1926–1929.

Darwin, Charles. 1871. *The descent of man* (1899, 2d ed.). London: John Murray.

Davis, Shelton H. 1993. Hard choices: Indigenous economic development and intellectual property rights. *Akwe:kon Journal,* Winter, pp. 19–25.

Davis, Wade. 1986. *The serpent and the rainbow.* New York: Simon and Schuster.

———. 1996. *One river: Science, adventure and hallucinogenics in the Amazon Basin.* New York: Simon and Schuster.

Dawkins, Richard. 1976. *The selfish gene*. Oxford: Oxford University Press.

———. 1982. *The extended phenotype: The gene as the unit of selection.* Oxford: W. H. Freeman.

———. 1986. *The blind watchmaker.* London: Penguin Books.

de Duve, Christian. 1984. *A guided tour of the living cell.* 2 vols. New York: Scientific American Library.

Delaby, Laurence. 1976. *Chamanes toungouses.* Paris: Université de Paris X, Centre d'Etudes Mongoles ("Etudes Mongoles . . . et Sibériennes"), volume 7.

Deliganis, Anna V., et al. 1991. Differential interactions of dimethyltryptamine (DMT) with $5-HT_{1A}$ and $5-HT_2$ receptors. *Biochemical Pharmacology,* 41(11):1739–1744.

Delsemme, Armand. 1994. *Les origines cosmiques de la vie: Du big bang à l'Homme.* Paris: Flammarion.

De Mille, Richard, ed. 1980. *The Don Juan papers: Further Castaneda controversies.* Santa Barbara: Ross-Erikson.

Deneris, Evan S., et al. 1991. Pharmacological and functional diversity of neuronal nicotinic acetylcholine receptors. *Trends in Pharmacological Science,* 12:34–40.

de Rosnay, Joël. 1966. *Les origines de la vie: De l'atome à la cellule.* Paris: Editions du Seuil.

Descola, Philippe. 1996. *The spears of twilight: Life and death in the Amazon jungle.* London: HarperCollins.

Devereux, George. 1956. Normal and abnormal: The key problem of psychiatric anthropology. In *Some uses of anthropology: Theoretical and applied,* The Anthropological Society of Washington, ed., pp. 23–48. Washington, D.C.: The Anthropological Society of Washington.

Diószegi, Vilmos. 1974. Shamanism. *Encyclopaedia Britannica,* 15th ed., vol. 16, pp. 638–641.

Dishotsky, Norman I., et al. 1971. LSD and genetic damage: Is LSD chromosome damaging, carcinogenic, mutagenic or teratogenic? *Science,* 172 (3982):431–440.

Dobkin de Rios, Marlene. 1972. *Visionary vine: Hallucinogenic healing in the Peruvian Amazon.* Prospect Heights, IL: Waveland Press.

———. 1973. Curing with ayahuasca in an urban slum. In *Hallucinogens and shamanism,* Michael Harner, ed., pp. 67–85. Oxford: Oxford University Press.

———. 1974. Cultural persona in drug-induced altered states of consciousness. In *Social and cultural identity*, T. K. Fitzgerald, ed., pp. 15–23. Athens: University of Georgia Press.

Dobkin de Rios, Marlene, and Fred Katz. 1975. Some relationships between music and hallucinogenic ritual: The "jungle gym" of consciousness. *Ethos*, 3(1):64–76.

Drummond, Lee. 1981. The serpent's children: Semiotics of cultural genesis in Arawak and Trobriand myth. *American Ethnologist*, 8(3):633–660.

Dubochet, Jacques. 1993. Twisting in a crowd. *Trends in Cell Biology*, 3(1):1–3.

Eisner, Thomas. 1990. Chemical prospecting: A proposal for action. In *Ecology, economics, ethics: The broken circle*, F. H. Bormann and S. R. Kellert, eds., pp. 196–202. New Haven: Yale University Press.

Eliade, Mircea. 1949. *Traité d'histoire des religions*. Paris: Payot.

———. 1951. *Le chamanisme et les techniques archaïques de l'extase*. Paris: Payot.

———. 1964. *Shamanism: Archaic techniques of ecstasy*. New York: Arkana.

———. 1972. *Religions australiennes*. Paris: Payot.

Elick, John W. 1969. An ethnography of the Pichis Valley Campa of Eastern Peru. Ph.D. dissertation, UCLA. Ann Arbor: University Microfilms.

Elisabetsky, Elaine. 1991. Folklore, tradition or know-how? *Cultural Survival Quarterly*, 15(3):9–13.

Evans, Gary D. 1993. Cigarette smoke = radiation hazard. *Pediatrics*, 92(3):464–465.

Evinger, Marian J., et al. 1994. A single transmitter regulates gene expression through two separate mechanisms: Cholinergic regulation of phenylethanolamine N-methyltransferase mRNA via nicotinic and muscarinic pathways. *Journal of Neuroscience*, 14(4):2106–2116.

Farin, Claus-Jürgen, et al. 1990. Mechanisms involved in the transcriptional activation of proenkephalin gene expression in bovine chromaffin cells. *Journal of Biology and Chemistry*, 265 (31):19116– 19121.

Farnsworth, Norman R. 1988. Screening plants for new medicines. In *National forum on biodiversity*, E. O. Wilson and F. M. Peters, eds., pp. 83–97. Washington, D.C.: National Academy Press.

Federal Chancellery of Switzerland. 1991. *Ordonnance sur les denrées alimentaires*. Berne: Chancellerie Fédérale.

Federal Office of Public Health. 1994. Le tabagisme en Suisse—quelques données. Berne: Office Fédéral de la Santé publique.

Flam, Faye. 1994. Hints of a language in junk DNA. *Science,* 266:1320.

Foucault, Michel. 1961. *Folie et déraison: Histoire de la folie à l'âge classique.* Paris: Plon.

Frank-Kamenetskii, Maxim D. 1993. *Unraveling DNA.* New York: VCH Publishers.

Fredrickson, James K., and Tullis C. Onstott. 1996. Microbes deep inside the earth. *Scientific American,* 275 (4):42–47.

Freedman, David H. 1994. Lone wave. *Discover,* 15(12):62–68.

Friedlander, Walter J. 1992. *The golden wand of medicine—a history of the caduceus symbol in medicine.* New York: Greenwood Press.

Furst, Peter T. 1994. Introduction: An overview of shamanism. In *Ancient traditions: Shamanism in Central Asia and the Americas,* Gary Seaman and Jane S. Day, eds., pp. 1–28. Niwot: University Press of Colorado.

Furst, Peter T., ed. 1972. *Flesh of the gods: The ritual use of hallucinogens.* New York: Praeger.

Galle, M., et al. 1991. Biophoton emission from *Daphnia magna*: A possible factor in the self-regulation of swarming. *Experientia,* 47 (5):457–460.

Gardiner, Alan. 1950. *Egyptian grammar: Being an introduction to the study of hieroglyphs.* London: Oxford University Press.

Garza, Mercedes de la. 1990. *Le chamanisme nahua et maya.* Paris: Trédaniel.

Gasché, Jürg. 1989–1990. A propos d'une nouvelle expérience d'éducation bilingue au Pérou. L'indigénisation d'un programme; sa critique de l'anthropologue. *Journal de la Société Suisse des Américanistes,* 53–54:131–142.

———. 1993. Elaboration et fonctions d'un discours anthropologique interculturel dans le cadre d'un programme d'éducation interculturelle bilingue au Pérou. *Bulletin de l'Association pour la Recherche Interculturelle (ARIC),* 22:23–51.

Gebhart-Sayer, Angelika. 1986. Una terapia estética: Los diseños visionarios del ayahuasca entre los Shipibo-Conibo. *América Indígena,* 46(1):189–218.

———. 1987. *Die Spitze des Bewusstseins: Untersuchungen zu Weltbild und Kunst der Shipibo-Conibo.* Hohenschäftlarn: Klaus Renner Verlag.

Geertz, Clifford. 1966. Religion as a cultural system. In *Anthropological approaches to the study of religion*, M. Banton et al., eds., pp. 1–46. London: A.S.A Monographs.

Gilbert, Walter. 1992. A vision of the grail. In *The code of codes: Scientific and social issues in the Human Genome Project*, Daniel J. Kevles and Leroy Hood, eds., pp. 83–97. Cambridge, MA: Harvard University Press.

Glennon, Richard A., et al. 1984. Evidence for 5-HT$_2$ involvement in the mechanism of action of hallucinogenic drugs. *Life Sciences*, 35:2505–2511.

Goodsell, David S. 1993. *The machinery of life.* New York: Springer-Verlag.

Goodwin, Brian. 1994. *How the leopard changed its spots.* London: Weidenfeld and Nicholson.

Graves, Robert. 1955. *The Greek myths.* London: Penguin Books.

Grinspoon, Lester, and James B. Bakalar. 1979. *Psychedelic drugs reconsidered.* New York: Basic Books.

Gu, Qiao. 1992. Quantum theory of biophoton emission. In *Recent advances in biophoton research and its applications*, F.-A. Popp et al., eds., pp. 59–112. Singapore: World Scientific.

Guénon, René. 1962. *Symboles fondamentaux de la science sacrée.* Paris: Gallimard.

Gurvich, Anna A. 1992. Mitogenic radiation as an evidence of nonequilibrium properties of living matter. In *Recent advances in biophoton research and its applications*, F.-A. Popp et al., eds., pp. 457–468. Singapore: World Scientific.

Halifax, Joan. 1979. *Shamanic voices: A survey of visionary narratives.* New York: E. P. Dutton.

Hall, Daniel B., et al. 1996. Oxidative DNA damage through long-range electron transfer. *Nature*, 382:731–735.

Hamayon, Roberte. 1978. Soigner la mort pour guérir le vif. *Nouvelle Revue de Psychanalyse*, 17:55–72.

———. 1982. Des chamanes au chamanisme. *L'Ethnographie*, 87–88(1–2):13–47.

———. 1990. *La chasse à l'âme: Esquisse d'une théorie du chamanisme à partir d'exemples sibériens.* Nanterre: Société d'Ethnologie.

Hare, E. H. 1973. A short note on pseudo-hallucinations. *British Journal of Psychiatry*, 122:469–476.

Harner, Michael. 1968. The sound of rushing water. *Natural History Magazine*, 77(6):28–33, 60–61.

————. 1973. Common themes in South American Indian yagé experiences. In *Hallucinogens and shamanism,* Michael Harner, ed., pp. 155–175. Oxford: Oxford University Press.

————. 1980. *The way of the shaman.* New York: Harper & Row.

Harner, Michael, ed. 1973. *Hallucinogens and shamanism.* Oxford: Oxford University Press.

Hayes, John M. 1996. The earliest memories of life on Earth. *Nature,* 384 (6604):21–22.

Heald, Rebecca, et al. 1996. Self-organization of microtubules into bipolar spindles around artificial chromosomes in *Xenopus* egg extracts. *Nature,* 382 (6590):420–425.

Hill, Jonathan D. 1992. A musical aesthetic of ritual curing in the Northwest Amazon. In *Portals of power: Shamanism in South America,* E. Jean Matteson Langdon and Gerhard Baer, eds., pp. 175–210. Albuquerque: University of New Mexico Press.

Hilts, Philip J. 1996. Unexplained holes remain in fuller genetic map. *New York Times,* 25 October, p. C19.

Hinkle, Gregory, et al. 1994. Phylogeny of the attine ant fungi based on analysis of small subunit ribosomal RNA gene sequences. *Science,* 266:1695–1697.

Ho, Mae-Wan, and Fritz-Albert Popp. 1993. Biological organization, coherence and light emission from living organisms. In *Thinking about biology,* W. D. Stein and F. J. Varela, eds., pp. 183–213. New York: Addison-Wesley.

Hoffer, Abraham, and Humphry Osmond. 1967. *The hallucinogens.* New York: Academic Press.

Hofmann, Albert. 1983. *LSD, my problem child.* Boston: Houghton Mifflin.

Holder, Nigel, and Andrew McMahon. 1996. Genes from zebrafish screens. *Nature,* 384:515–516.

Hölldobler, Bert, and Edward O. Wilson. 1990. *The ants.* Cambridge, MA: Harvard University Press.

Hoppál, Mihály. 1987. Shamanism: An archaic and/or recent system of beliefs. In *Shamanism,* Shirley Nicholson, ed., pp. 76–100. London: Quest.

Horgan, John. 1994. Can science explain consciousness? *Scientific American,* 271(1):88–94.

Huffman, Michael A. 1995. La pharmacopée des chimpanzées. *La Recherche,* 280:66–71.

Hultkrantz, Ake. 1978. Ecological and phenomenological aspects of shamanism. In *Studies in Lapp shamanism,* Louise Bäckman and Ake

Hultkrantz, eds., pp. 9–35. Stockholm: Almqvist and Wiksell International.

Hunt, Gavin R. 1996. Manufacture and use of hook-tools by New Caledonian crows. *Nature,* 379 (6562):249–250.

Huxley, Francis. 1974. *The way of the sacred.* Garden City, NY: Doubleday and Company.

Hyams, Jeremy. 1996. Look Ma, no chromosomes! *Nature,* 382 (6590):397–398.

Illius, Bruno. 1992. The concept of nihue among the Shipibo-Conibo of Eastern Peru. In *Portals of power: Shamanism in South America,* E. Jean Matteson Langdon and Gerhard Baer, eds., pp. 63–77. Albuquerque: University of New Mexico Press.

Iversen, S. D., and L. L. Iversen. 1981. *Behavioural pharmacology.* Oxford: Oxford University Press.

Jacob, François. 1974. *La logique du vivant: Une histoire de l'hérédité.* Paris: Gallimard.

Jacq, Christian. 1993. *La vallée des Rois: Images et mystères.* Paris: Perrin.

———. 1994. *Le petit Champollion illustré: Les hiéroglyphes à la portée de tous, ou Comment devenir scribe amateur tout en s'amusant.* Paris: Robert Laffont.

Jakobson, Roman. 1973. *Essais de linguistique générale.* Vol. 2: *Rapports internes et externes du langage.* Paris: Les Editions de Minuit.

Johnson, George. 1994. Learning just how little is known about the brain. *New York Times,* 23 October, p. E5.

Jones, Steve. 1993. *The language of the genes.* London: Flamingo.

Judson, Horace F. 1992. A history of the science and technology behind gene mapping and sequencing. In *The code of codes: Scientific and social issues in the Human Genome Project,* Daniel J. Kevles and Leroy Hood, eds., pp. 37–80. Cambridge, MA: Harvard University Press.

Jung, Carl J., Charles Kerényi, and Paul Radin. 1958. *Le fripon divin.* Geneva: Georg Editeur.

Kahn, Charles H. 1979. *The art and thought of Heraclitus: An edition of the fragments with translation and commentary.* Cambridge: Cambridge University Press.

Kaplan, J., et al. 1974. Blood and urine levels of N,N-dimethyltryptamine following administration of psychoactive doses to human subjects. *Psychopharmacologia,* 38:239–245.

Kato, Takashi, and Lissy F. Jarvik. 1969. LSD-25 and genetic damage. *Diseases of the Nervous System*, 30:42–46.

Kato, Takashi, et al. 1970. Chromosome studies in pregnant rhesus macaques given LSD-25. *Diseases of the Nervous System*, 31:245–250.

Kauffman, Stuart. 1996. *Even peptides do it*. Nature, 382 (6591): 496–497.

Kensinger, Kenneth M. 1973. Banisteriopsis usage among the Peruvian Cashinahua. In *Hallucinogens and shamanism*, Michael Harner, ed., pp. 9–14. Oxford: Oxford University Press.

Keup, Wolfram, ed. 1970. *Origin and mechanisms of hallucinations*. New York: Plenum Press.

Kevles, Daniel J., and Leroy Hood, eds. 1992. *The code of codes: Scientific and social issues in the Human Genome Project*. Cambridge, MA: Harvard University Press.

King, Steven R. 1991. The source of our cures. *Cultural Survival Quarterly*, 15(3):19–22.

Klaassen, Curtis D., and King Lit Wong. 1993. Poisons and poisoning. In *Encyclopaedia Britannica*, 15th ed., vol. 25, pp. 908–929.

Kloppenburg, Jack, Jr. 1991. No hunting! Biodiversity, indigenous rights, and scientific poaching. *Cultural Survival Quarterly*, 15(3):14–18.

Koch-Grünberg, Theodor. 1917. *Vom Roroíma zum Orinoco: Ergebnisse einer Reise in Nordbrasilien und Venezuela in den Jahren, 1911–1913*. Vol. 2. Stuttgart: Strecker und Schröder.

Koistinaho, J., et al. 1993. Differential expression of immediate early genes in the superior cervical ganglion after nicotine treatment. *Neuroscience*, 56(3):729–739.

Kracke, Waud H. 1992. He who dreams: The nocturnal source of transforming power in Kagwahiv shamanism. In *Portals of power: Shamanism in South America*, E. Jean Matteson Langdon and Gerhard Baer, eds., pp. 127–148. Albuquerque: University of New Mexico Press.

Krajick, Kevin. 1997. The floating zoo. *Discover*, 18(2):66–73.

Kräupl Taylor, F. 1981. On pseudo-hallucinations. *Psychological Medicine*, 11:265–271.

Kuper, Adam. 1988. *The invention of primitive society: Transformations of an illusion*. London: Routledge.

La Barre, Weston. 1976. Stinging criticism from the author of *The Peyote Cult*. In *Seeing Castaneda*, Daniel C. Noel, ed., pp. 40–42. New York: G. P. Putnam's Sons.

Lamb, Bruce F. 1971. *Wizard of the Upper Amazon: The story of Manuel Córdova Rios.* Boston: Houghton Mifflin.

———. 1985. *Rio Tigre and beyond: The Amazon jungle medicine of Manuel Córdova.* Berkeley: North Atlantic Books.

Langaney, André. 1997. Ce que l'on ne sait pas de l'évolution. *La Recherche,* 296:118–124.

Langdon, E. Jean Matteson, and Gerhard Baer, eds. 1992. *Portals of power: Shamanism in South America.* Albuquerque: University of New Mexico Press.

Laszlo, Pierre. 1997. Origine de la vie: 100 000 milliards de scénarios. *La Recherche,* 296:26–30.

Leary, Timothy. 1966. Programmed communication during experience with DMT. *Psychedelic Review,* 8:83–95.

Lee, David H., et al. 1996. A self-replicating peptide. *Nature,* 382 (6591):525–528.

Lee, Martin A., and Bruce Shlain. 1985. *Acid dreams: The CIA, LSD and the sixties rebellion.* New York: Grove Press.

Lévi-Strauss, Claude. 1949a. *Les structures élémentaires de la parenté.* Paris: Presses Universitaires de France. (Published in English in 1969 as *The elementary structures of kinship.* London: Eyre and Spottiswoode.)

———. 1949b. L'efficacité symbolique. *Revue de l'Histoire des Religions,* 135 (1):5–27. (Published in Lévi-Strauss 1963a, pp. 186–205.)

———. 1950. The use of wild plants in tropical South America. In *Handbook of South American Indians,* Julian Steward, ed., vol. 6, pp. 465–486. Washington, D.C.: U.S. Government Printing Office, Bureau of American Ethnology.

———. 1955. *Tristes tropiques.* Paris: Terre Humaine/Poche.

———. 1963a. *Structural anthropology 1.* New York: Basic Books.

———. 1963b. *Totemism.* Boston: Beacon.

———. 1983. *Le regard éloigné.* Paris: Plon.

———. 1991a. Un entretien avec Claude Lévi-Strauss. *Le Monde,* 8 October, p. 2.

———. 1991b. *Histoire de lynx.* Paris: Plon.

Lewis, Diane. 1973. Anthropology and colonialism. *Current Anthropology,* 14:581–602.

Lewis, Ioan M. 1971. *Ecstatic religion: An anthropological study of spirit possession and shamanism.* London: Penguin Books.

Lewis, James A., et al. 1987. Cholinergic receptor mutants of the nematode *Caenorhabditis elegans*. *Journal of Neuroscience*, 7(10):3059–3071.

Lewontin, R. C. 1992. *The doctrine of DNA: Biology as ideology*. London: Penguin Books.

Li, Ke-hsueh. 1992. Coherence in physics and biology. In *Recent advances in biophoton research and its applications*, F. A. Popp et al., eds., pp. 157–195. Singapore: World Scientific.

Lipkin, R. 1994. Watching polymers wend their way along. *Science News*, 145 (May 7):293.

Lot-Falck, Eveline. 1963. A propos du terme chamane. In *Etudes Mongoles . . . et Sibériennes, 1977*, vol. 8, pp. 7–18.

———. 1973. Le chamanisme en Sibérie: Essai de mise au point. *Bulletin de l'Asie du Sud-Est et Monde Insulindien*, 4(3), pp. 1–10.

Luisi, Pier Luigi. 1993. Defining the transition to life: Self-replicating bounded structures and chemical autopoiesis. In *Thinking about biology*, W. D. Stein and F. J. Varela, eds., pp. 17–39. New York: Addison-Wesley.

Luna, Luis Eduardo. 1984. The concept of plants as teachers among four Peruvian shamans of Iquitos, Northeast Peru. *Journal of Ethnopharmacology*, 11:135–156.

———. 1986. *Vegetalismo: Shamanism among the mestizo population of the Peruvian Amazon*. Stockholm: Almqvist and Wiksell.

———. 1992. Magic melodies among the mestizo shamans of the Peruvian Amazon. In *Portals of power: Shamanism in South America*, E. Jean Matteson Langdon and Gerhard Baer, eds., pp. 231–253. Albuquerque: University of New Mexico Press.

Luna, Luis Eduardo, and Pablo Amaringo. 1991. *Ayahuasca visions: The religious iconography of a Peruvian shaman*. Berkeley: North Atlantic Books.

Mabit, Jacques Michel. 1988. *L'hallucination par l'ayahuasca chez les guérisseurs de la Haute-Amazone péruvienne (Tarapoto)*. Lima: Institut Français d'Etudes Andines.

Mabit, Jacques Michel, et al. 1992. Consideraciones acerca del brebaje ayahuasca y perspectivas terapeuticas. *Revista de Neuro-Psiquiatria*, 55(2):118–131 (Lima, Peru).

Maier, Michael. 1965. *Atalante fugitive* (originally published 1618). Paris: Librairie de Médicis.

Malinowski, Bronislaw. 1922. *Argonauts of the Western Pacific*. New York: E. P. Dutton (1961 edition).

Mann, John. 1992. *Murder, magic and medicine*. Oxford: Oxford University Press.

Margulis, Lynn, and Dorion Sagan. 1986. *Microcosmos: Four billion years of microbial evolution*. New York: Touchstone. (Published in French with modifications as *L'univers bactériel: Les nouveaux rapports de l'homme et de la nature*. Paris: Albin Michel, 1989.)

Martell, Edward A. 1982. Radioactivity in cigarette smoke. *New England Journal of Medicine*, 307:309–310.

Mayr, Ernst F. 1982. *The growth of biological thought: Diversity, evolution, and inheritance*. Cambridge, MA: Belknap Press, Harvard University.

———. 1983. How to carry out the adaptationist program. *American Naturalist*, 121:324–334.

———. 1988. *Toward a new philosophy of biology: Observations of an evolutionist*. Cambridge, MA: Harvard University Press.

McGinn, Colin. 1994. How I may or may not have solved the mind-body problem and nearly ruined my life. *Lingua Franca*, November/December, pp. 67–71.

McKenna, Dennis J., and Terence McKenna. 1975. *The invisible landscape: mind, hallucinogens and the* I Ching. New York: Seabury Press.

McKenna, Dennis J., and Stephen J. Peroutka. 1990. Neurochemistry and neurotoxicity of 3,4-methylenedioxymethamphetamine (MDMA, "Ecstasy"). *Journal of Neurochemistry*, 54(1):14–22.

McKenna, Dennis J., L. E. Luna, and G. H. N. Towers. 1986. Ingredientes biodinámicos en las plantas que se mezclan al ayahuasca: Una farmacopea tradicional no investigada. *América Indígena*, 46(1): 73–99.

McKenna, Dennis J., G. H. N. Towers, and F. Abbott. 1984. Monoamine oxidase inhibitors in South American hallucinogenic plants: Tryptamine and beta-carboline constituents of ayahuasca. *Journal of Ethnopharmacology*, 10:195–223.

McKenna, Dennis J., et al. 1989. Common receptors for hallucinogens in rat brain: A comparative autoradiographic study using [^{125}I]LSD and [^{125}I]DOI, a new psychotomimetic radioligand. *Brain Research*, 476:45–56.

McKenna, Terence. 1988. *The ethnobotany of shamanism* (oral presentation, 6 cassettes). Mill Valley, CA: Sound Photosynthesis.

———. 1991. *The archaic revival: Speculations on psychedelic mushrooms, the Amazon, virtual reality, UFOs, evolution, shamanism, the rebirth of the Goddess and the end of history*. San Francisco: Harper-Collins.

————. 1993. *True hallucinations: Being an account of the author's extraordinary adventures in the devil's paradise*. San Francisco: HarperCollins.

Mei, Wei-ping. 1992. Ultraweak photon emission from synchronized yeast *(Saccharomyces cerevisiae)* as a function of the cell division cycle. In *Recent advances in biophoton research and its applications*, F.-A. Popp et al., eds., pp. 243–258. Singapore: World Scientific.

Métraux, Alfred. 1946. Twin heroes in South American mythology. *Journal of American Folklore*, 59:114–123.

————. 1967. *Religions et magies indiennes d'Amérique du Sud*. Paris: Gallimard.

Mitchell, S. N., et al. 1993. Increases in tyrosine hydroxylase messenger RNA in the locus coeruleus after a single dose of nicotine are followed by time-dependent increases in enzyme activity and noradrenaline release. *Neuroscience*, 56(4):989–997.

Mitriani, Philippe. 1982. Aperçu critique des approches psychiatriques du chamanisme. *L'Ethnographie*, 87–88(1–2):241–258.

Mojzsis, S. J., et al. 1996. Evidence for life on Earth before 3,800 million years ago. *Nature*, 384 (6604):55–59.

Monod, Jacques. 1971. *Chance and necessity: An essay on the natural philosophy of modern biology*. London: Founts Paperback.

Moore, Melissa J. 1996. When the junk isn't junk. *Nature*, 379:402–403.

Moorhead, Paul S., and Martin M. Kaplan, eds. 1967. *Mathematical challenges to the neo-Darwinian interpretation of evolution*. Philadelphia: Wistar Institute Press.

Morowitz, Harold J. 1985. *Mayonnaise and the origins of life*. New York: Scribner's.

Mullis, Kary. 1994. The polymerase chain reaction (Nobel lecture). *Angewandte Chemie International Edition in English*, 33:1209–1213.

Mundkur, Balaji. 1983. *The cult of the serpent: An interdisciplinary survey of its manifestations and origins*. Albany: State University of New York Press.

Murphy, C.-J., et al. 1993. Long-range photoinduced electron transfer through a DNA helix. *Science*, 262(5136):1025–1029.

Mushegian, Arcady R., and Eugene V. Koonin. 1996. A minimal gene set for cellular life derived by comparison of complete bacterial genomes. *Proceedings of the National Academy of Sciences (U.S.)*, 93:10268–10273.

Naranjo, Plutarco. 1986. El ayahuasca en la arqueologia ecuatoriana. *América Indígena*, 46(1):117–127.

Narby, Jeremy. 1986. El Banco Agrario y las comunidades Asháninka del Pichis: Crédito promocional para productores nativos. *Amazonia Indígena*, 6(12):14–21.

———. 1989. Visions of land: The Ashaninca and resource development in the Pichis Valley in the Peruvian central jungle. Ph.D. dissertation, Stanford University. Ann Arbor: University Microfilms.

———. 1990. *Amazonie, l'espoir est indien*. Paris: Favre.

Nash, Madeleine J. 1995. When life exploded. *Time*, 5 December, pp. 66–74.

Niggli, Hugo J. 1992. Biophoton re-emission studies in carcinogenic mouse melanoma cells. In *Recent advances in biophoton research and its applications*, F.-A. Popp et al., eds., pp. 231–242. Singapore: World Scientific.

Noel, Daniel C., ed. 1976. *Seeing Castaneda*. New York: G. P. Putnam's Sons.

Noll, Richard. 1983. Shamanism and schizophrenia: A state-specific approach to the "schizophrenia metaphor" of shamanic states. *American Ethnologist*, 10:443–459.

Nowak, Rachel. 1994. Mining treasures from "junk DNA." *Science*, 263:608–610.

Orgel, Leslie E., and Francis H. C. Crick. 1980. Selfish DNA: The ultimate parasite. *Nature*, 284:604–607.

Pang, Ying, et al. 1993. Acute nicotine injections induce c-*fos* mostly in non-dopaminergic neurons of the midbrain of the rat. *Molecular Brain Research*, 20(1–2):162–170.

Pennisi, Elizabeth. 1994. Does nonsense DNA speak its own dialect? *Science News*, 146(24):391.

Penrose, Roger. 1994. *Shadows of the mind: A search for the missing science of consciousness*. Oxford: Oxford University Press.

Perrin, Michel. 1992a. The body of the Guajiro shaman: Symptoms or symbols. In *Portals of power: Shamanism in South America*, E. Jean Matteson Langdon and Gerhard Baer, eds., pp. 103–125. Albuquerque: University of New Mexico Press.

———. 1992b. *Les praticiens du rêve: Un exemple de chamanisme*. Paris: Presses Universitaires de France.

Perry, Nicolette. 1983. *Symbiosis: Close encounters of the natural kind*. Poole, Dorset: Blandford Press.

Piaget, Jean. 1975. *The development of thought*. Oxford: Blackwell.

Pierce, Pamela A., and Stephen J. Peroutka. 1989. Hallucinogenic drug interactions with neurotransmitter receptor binding sites in human cortex. *Psychopharmacology,* 97:118–122.

Pitt, Bruce R., et al. 1994. Serotonin increases DNA synthesis in rat proximal and distal pulmonary vascular smooth muscle cells in culture. *American Journal of Physiology,* 266: L178–L186.

Plotkin, Mark. 1993. *Tales of a shaman's apprentice: An ethnobotanist searches for new medicines in the Amazon rainforest.* New York: Viking Press.

Pollack, Robert. 1994. *Signs of life: The language and meanings of DNA.* New York: Viking.

———. 1997. A crisis in scientific morale. *Nature,* 385:673–674.

Popp, Fritz-Albert. 1986. On the coherence of ultraweak photon emission from living tissues. In *Disequilibrium and self-organization,* C. W. Kilmister, ed., pp. 207–230. Dordrecht: Reidel.

———. 1992a. Some essential questions of biophoton research and probable answers. In *Recent advances in biophoton research and its applications,* F.-A. Popp et al., eds., pp. 1–46. Singapore: World Scientific.

———. 1992b. Evolution as the expansion of coherent states. In *Recent advances in biophoton research and its applications,* F.-A. Popp et al., eds., pp. 445–456. Singapore: World Scientific.

Popp, Fritz-Albert, Qiao Gu, and Ke-Hsueh Li. 1994. Biophoton emission: Experimental background and theoretical approaches. *Modern Physics Letters B,* 8(21–22):1269–1296.

Popp, Fritz-Albert, et al., eds. 1992. *Recent advances in biophoton research and its applications.* Singapore: World Scientific.

Popper, Karl. 1974. *Unended quest: An intellectual autobiography.* London: Fontana.

Posey, Darrell A. 1990. Intellectual property rights and just compensation for indigenous knowledge. *Anthropology Today,* 6(4): 13–16.

———. 1991. Effecting international change. *Cultural Survival Quarterly,* 15(3):29–35.

———. 1994. International agreements and intellectual property right protection for indigenous people. In *Intellectual property rights for indigenous peoples: A sourcebook,* Tom Greaves, ed., pp. 223–243. Oklahoma City: Society for Applied Anthropology.

Pratt, Mary Louise. 1986. Fieldwork in common places. In *Writing culture: The poetics and politics of ethnography,* James Clifford and George E. Marcus, eds., pp. 27–50. Berkeley: University of California Press.

Radman, Miroslav, and Robert Wagner. 1988. The high fidelity of DNA duplication. *Scientific American*, 259(8):24–30.

Rattemeyer, M., et al. 1981. Evidence of photon emission from DNA in living systems. *Naturwissenschaften*, 68:572–573.

Reichel-Dolmatoff, Gerardo. 1971. *Amazonian cosmos: The sexual and religious symbolism of the Tukano Indians*. Chicago: University of Chicago Press.

———. 1972. "The cultural context of an aboriginal hallucinogen: *Banisteriopsis caapi.*" In *Flesh of the Gods: the ritual use of hallucinogens*, Peter T. Furst, ed., pp. 84–113. New York: Praeger.

———. 1975. *The shaman and the jaguar: A study of narcotic drugs among the Indians of Colombia*. Philadelphia: Temple University Press.

———. 1978. *Beyond the Milky Way: Hallucinatory imagery of the Tukano Indians*. Los Angeles: UCLA Latin America Center.

———. 1979. Desana shaman's rock crystals and the hexagonal universe. *Journal of Latin American Lore*, 5(1):117–128.

———. 1981. Brain and mind in Desana shamanism. *Journal of Latin American Lore*, 7(1):73–98.

———. 1987. The Great Mother and the Kogi universe: A concise overview. *Journal of Latin American Lore*, 13 (1):73–113.

Reisse, Jacques. 1988. Origine de la vie. In *Les origines*, Yves Coppens et al., eds., pp. 96–118. Paris: L'Harmattan.

Renard-Casevitz, France-Marie. 1993. Guerriers du sel, sauniers de la paix. *L'Homme*, 126–128(2–4):25–43.

Renck, Jean-Luc. 1989. Comportement vocal et communication chez le chat domestique (*Felis silvestris catus*). Mémoire de licence, Institut de zoologie, Université de Neuchâtel (unpublished manuscript).

Rivier, Laurent, and Jan-Erik Lindgren. 1972. "Ayahuasca," the South American hallucinogenic drink: An ethnobotanical and chemical investigation. *Economic Botany*, 26:101–129.

Roe, Peter G. 1982. *The cosmic zygote: Cosmology in the Amazon Basin*. New Brunswick, NJ: Rutgers University Press.

Rosaldo, Renato. 1980. Doing oral history. *Social Analysis*, 4:89–99.

———. 1989. *Culture and truth: The remaking of social analysis*. Boston: Beacon Press.

Rosenberg, D. E., et al. 1963. Comparison of a placebo, N-dimethyltryptamine and 6-hydroxy-N-dimethyltryptamine in man. *Psychopharmacologia*, 4:39–42.

Rouget, Gilbert. 1980. *La musique et la transe: Esquisse d'une théorie générale des relations de la musique et de la possession.* Paris: Gallimard.

Rouhi, A. Maureen. 1997. Seeking drugs in natural products. *Chemical and Engineering News,* 75 (14):14–29.

Sagan, Carl, and the Editors. 1993. Life. *Encyclopaedia Britannica,* 15th ed., vol. 22, pp. 964–981.

Saïd, Edward W. 1978. *Orientalism.* New York: Pantheon Books.

Sai-Halasz, A., et al. 1958. Dimethyltryptamin: Ein neues Psychoticum. *Psychiatria et Neurologia (Basel),* 135:285–301.

Sankarapandi, S. 1994. Cracking the cancer code. *Down to Earth,* 15 September, pp. 25–31.

Schiefelbein, Susan. 1986. Le début du voyage. In *L'admirable machine humaine,* National Geographic Society, ed., pp. 13–53. Paris: Editions du Chêne.

Schultes, Richard Evans. 1969. Hallucinogens of plant origin. *Science,* 163(864): 245–254.

———. 1972. An overview of hallucinogens in the Western Hemisphere. In *Flesh of the Gods: The ritual use of hallucinogens,* Peter T. Furst, ed., pp. 3–54. New York: Praeger.

Schultes, Richard Evans, and Albert Hofmann. 1979. *Plants of the gods: Origins of hallucinogenic use.* London: McGraw-Hill.

———. 1980. *The botany and chemistry of hallucinogens,* 2d ed. Springfield, IL: Charles Thomas.

Schultes, Richard Evans, and Robert F. Raffauf. 1990. *The healing forest: Medicinal and toxic plants of the Northwest Amazonia.* Portland, OR: Dioscorides Press.

———. 1992. *Vine of the soul: Medicine men, their plants and rituals in the Colombian Amazonia.* Oracle, AZ: Synergetic Press.

Schultes, Richard Evans, and Siri von Reis, eds. 1995. *Ethnobotany: Evolution of a discipline.* New York: Chapman and Hall.

Schützenberger, Marcel-Paul. 1996. Les failles du darwinisme. *La Recherche,* 283:87–90.

Seaman, Gary, and Jane S. Day, eds. 1994. *Ancient traditions: Shamanism in Central Asia and the Americas.* Niwot, CO: University Press of Colorado.

Shanon, Benny. 1978. The genetic code and human language. *Synthese,* 39:401–415.

Shapiro, James A. 1996. In the details ... what? *National Review,* 16 September, pp. 62–65.

Shapiro, Robert. 1986. *Origins: A skeptic's guide to the creation of life on Earth.* London: Heinemann.

———. 1988. *L'origine de la vie.* Paris: Flammarion.

———. 1994a. Préface de la nouvelle édition française. In *L'origine de la vie*, 2d ed., pp. I–III. Paris: Flammarion.

———. 1994b. Epilogue: Barcelona 1993. In *L'origine de la vie*, 2d ed., pp. 415–426. Paris: Flammarion.

Shulgin, Alexander T. 1992. *Controlled substances: Chemical and legal guides to federal drug laws.* Berkeley: Ronin Publishing.

Siegel, Ronald K., and Murray E. Jarvik. 1975. Drug-induced hallucinations in animals and man. In *Hallucinations: Behavior, experience, and theory*, R. K. Siegel and L. J. West, eds., pp. 81–162. New York: Wiley.

Silverman, Julian. 1967. Shamans and acute schizophrenia. *American Anthropologist*, 69:21–31.

Siskind, Janet. 1973. Visions and cures among the Sharanahua. In *Hallucinogens and shamanism*, Michael Harner, ed., pp. 28–39. Oxford: Oxford University Press.

Slade, Peter. 1976. Hallucinations. *Psychological Medicine*, 6:7–13.

Slade, Peter D., and Richard P. Bentall. 1988. *Sensory deception: A scientific analysis of hallucination.* London: Croom Helm.

Smith, Douglas L. 1994. Sing a song of benzene, a pocket full of π. *Engineering and Science*, 58(1):27–38.

Smith, Richard Chase. 1982. *The dialectics of domination in Peru: Native communities and the myth of the vast Amazonian emptiness.* Cambridge, MA: Cultural Survival.

Smythies, John R. 1970. The chemical nature of the receptor site: A study in the stereochemistry of synaptic mechanisms. *International Review of Neurobiology*, 13:181–222.

Smythies, John R., and F. Antun. 1969. Binding of tryptamine and allied compounds to nucleic acids. *Nature*, 223:1061–1063.

Smythies, John R., et al. 1979. Identification of dimethyltryptamine and O-methylbufotenin in human cerebrospinal fluid by combined gas chromotography/mass spectrometry. *Recent Advances in Biological Psychiatry*, 14:549–556.

Snyder, Solomon H. 1986. *Drugs and the brain.* New York: Scientific American Library.

Stafford, Peter. 1977. *Psychedelics encyclopedia.* Berkeley: And/Or Press.

——. 1992. *Psychedelics encyclopedia,* 3d ed. Berkeley: Ronin Publishing.

Stein, W. D., and F. J. Varela, eds. 1993. *Thinking about biology.* New York: Addison-Wesley.

Stocco, Patrick. 1994. *Génie génétique et environnement: Principes fondamentaux et introduction à la problématique.* Geneva: Georg Editeur.

Strassman, Rick J. 1991. Human hallucinogenic drug research in the United States: A present-day case history and review of the process. *Journal of Psychoactive Drugs,* 23(1):29–38.

Strassman, Rick J., and Clifford R. Qualls. 1994. Dose-response study of N,N-dimethyltryptamine in humans. I. Neuroendocrine, autonomic and cardiovascular effects. *Archives of General Psychiatry,* 51:85–97.

Strassman, Rick J., et al. 1994. Dose-response study of N,N-dimethyltryptamine in humans. II. Subjective effects and preliminary results of a new rating scale. *Archives of General Psychiatry,* 51:98–108.

Sullivan, Lawrence E. 1988. *Icanchu's drum: An orientation to meaning in South American religions.* New York: Macmillan.

Swenson, Sally, and Jeremy Narby. 1985. Poco a poco, cual si fuera un tornillo: El Programa de Integración Indígena del Pichis. *Amazonia Indígena,* 5(10):17–26.

——. 1986. The Pichis-Palcazu Special Project in Peru—a consortium of international lenders. *Cultural Survival Quarterly,* 10(1):19–24.

Szára, S. 1956. Dimethyltryptamine—its metabolism in man; the relation of its psychotic effect to the serotonin metabolism. *Experientia (Basel),* 12:441–442.

——. 1957. The comparison of the psychotic effect of tryptamine derivatives with the effects of mescaline and LSD-25 in self-experiments. In *Psychotropic drugs,* S. Garattini and V. Ghetti, eds., pp. 460–467. Amsterdam: Elsevier Publishing.

——. 1970. DMT (N,N-dimethyltryptamine) and homologues: Clinical and pharmacological considerations. In *Psychotomimetic drugs,* D. H. Efron, ed., pp. 275–286. New York: Raven Press.

Taussig, Michael. 1987. *Shamanism, colonialism, and the wild man: A study in terror and healing.* Chicago: University of Chicago Press.

——. 1989. The nervous system: Homesickness and Dada. *Stanford Humanities Review,* 1(1):44–81.

——. 1992. *The nervous system.* New York: Routledge.

Thiébot, Marie-Hélène, and Michel Hamon. 1996. Un agent multiple: la sérotonine. *Pour la Science,* 221:82–89.

Thuillier, Pierre. 1986. Du rêve à la science: Le serpent de Kekulé. *La Recherche,* 17 (175):386–390.

Townsley, Graham. 1993. Song paths: The ways and means of Yaminahua shamanic knowledge. *L'Homme,* 126–128(2–4):449–468.

Trémolières, Antoine. 1994. *La vie plus têtue que les étoiles.* Paris: Nathan.

Trinh T. Minh-ha. 1989. *Woman, native, other: Writing postcoloniality and feminism.* Bloomington: Indiana University Press.

Tsing, Anna Lowenhaupt. 1993. *In the realm of the Diamond Queen: Marginality in an out-of-the-way place.* Princeton: Princeton University Press.

Tylor, Edward B. 1866. The religion of savages. *The Fortnightly Review,* 6:71–86.

Van de Kar, Louis. 1991. Neuroendocrine pharmacology of serotonergic (5-HT) neurons. *Annual Review of Pharmacology and Toxicology,* 31:289–320.

Van Gennep, Arnold. 1903. De l'emploi du mot "chamanisme." *Revue de l'Histoire des Religions,* 47(1):51–57.

Van Wijk, Roeland, and Hans van Aken. 1992. Spontaneous and light-induced photon emission by rat hepatocytes and by hepatoma cells. In *Recent advances in biophoton research and its applications,* F.-A. Popp et al., eds., pp. 207–229. Singapore: World Scientific.

Varese, Stefano. 1973. *La sal de los cerros,* 2d ed. Lima: Retablo de Papel Ediciones.

Wade, Nicholas. 1995a. Double helixes, chickens and eggs. *New York Times Magazine,* 29 January, p. 20.

———. 1995b. Rapid gains are reported on genome. *New York Times,* 28 September, p. A13.

———. 1995c. Are we aliens? *New York Times Magazine,* 22 November, pp. 22–23.

———. 1997a. Big stride for researchers in human-gene mapping. *New York Times,* 15 March, p. 1.

———. 1997b. Thinking small pays off big in gene quest. *New York Times,* 2 February, pp. A1, A14.

———. 1997c. Now playing at a nearby lab: "Revenge of the Fly People." *New York Times,* 20 May, pp. B7, B12.

Wagner, T. E. 1969. In vitro interaction of LSD with purified calf thymus DNA. *Nature,* 222:1170–1172.

Wan, D. C., et al. 1991. Coordinate and differential regulation of proenkephalin A and PNMT mRNA expression in cultural bovine adrenal chromaffin cells: Responses to secretory stimuli. *Molecular Brain Research*, 9(1–2):103–111.

Watson, James D. 1968. *The double helix: A personal account of the discovery of the structure of DNA.* London: Weidenfeld and Nicolson.

Watson, James D., et al. 1987. *Molecular biology of the gene,* 4th ed. Menlo Park, CA: Benjamin/Cummings Publishing.

Weiss, Gerald. 1969. *The cosmology of the Campa Indians of Eastern Peru.* Ann Arbor: University Microfilms.

———. 1973. Shamanism and priesthood in light of the Campa ayahuasca ceremony. In *Hallucinogens and shamanism,* Michael Harner, ed., pp. 40–52. Oxford: Oxford University Press.

Wesson, Robert. 1991. *Beyond natural selection.* Cambridge, MA: MIT Press.

Whitten, Norman E. 1976. *Sacha Runa: Ethnicity and adaptation of Ecuadorian jungle Quichua.* Urbana: University of Illinois Press.

Wilbert, Johannes. 1972. Tobacco and shamanistic ecstasy among the Warao Indians of Venezuela. In *Flesh of the gods: The ritual use of hallucinogens,* Peter T. Furst, ed., pp. 55–83. New York: Praeger.

———. 1987. *Tobacco and shamanism in South America.* New Haven: Yale University Press.

Wilbert, Werner. 1996. Environment, society, and disease: The response of phytotherapy to disease among the Warao Indians of the Orinoco Delta. In *Medicinal resources of the tropical forest: Biodiversity and its importance to human health,* Michael J. Balick et al., eds., pp. 366–385. New York: Columbia University Press.

Wills, Christopher. 1989. *The wisdom of the genes: New pathways in evolution.* Oxford: Oxford University Press.

———. 1991. Exons, introns, and talking genes: The science behind the Human Genome Project. Oxford: Oxford University Press.

Wilson, Edward O. 1984. *Biophilia.* Cambridge, MA: Harvard University Press.

———. 1990. Biodiversity, prosperity, and value. In *Ecology, economics, ethics: The broken circle,* F. H. Bormann and S. R. Kellert, eds., pp. 3–10. New Haven: Yale University Press.

———. 1992. *The diversity of life.* New York: Penguin.

Wilson, Edward O., and F. M. Peter, eds. 1988. *National forum on biodiversity.* Washington, D.C.: National Academy Press.

Wright, Pablo G. 1992. Dream, shamanism, and power among the Toba of Formosa province. In *Portals of power: Shamanism in South America*, E. Jean Matteson Langdon and Gerhard Baer, eds., pp. 149–172. Albuquerque: University of New Mexico Press.

Yielding, K. Lemone, and Helene Sterglanz. 1968. Lysergic acid diethylamide (LSD) binding to deoxyribonucleic acid (DNA). *Proceedings of the Society for Experimental Biology and Medicine,* 128:1096–1098.

Yoon, Carol Kaesuk. 1995. Gene is fuse that drives the flower. *New York Times,* 17 October, p. B5.

Zhang, Dahong, and R. Bruce Nicklas. 1996. "Anaphase" and cytokinesis in the absence of chromosomes. *Nature,* 382 (6590):466–468.

ACKNOWLEDGMENTS

MANY THANKS TO THE FOLLOWING: first reader, Rachel Vuillaume Narby; research assistant, Marie-Claire Chappuis; third base coach, Jon Christensen; unconditional support, Willy Randin/Nouvelle Planète; epistemology, Suren Erkman; anthropology, Jürg Gasché; metaphysics, Richard Chappuis; biology, Jean-Luc Renck, Véronique Servais; botany, Mathias Läubli, Michel Mettraux; medicine, Gilbert Guignard; professional guidance, Henri Weissenbach; argumentation, Jeremy P. Tarcher; French language consultant, Fabienne Radi Maitre; images, ric@act; original readers, Christophe Berdat, Philippe Randin, Yona Birker Chavanne, Patrick Lyndon, Claude Corboz, Laurent Grand, Jacques Falquet, Jean-Pierre Hurni, Jacques Mabit, Jacob Granatouroff; English manuscript readers, Michael Harner, Patrick Lyndon, Adrian Franklin, Rob La Frenais, Philip Colchin, Francis Huxley; literary agent, Barbara Moulton; my professors, Humphry Osmond, Sylvia Yanagisako, Renato Rosaldo, Shelton Davis, Stefano Varese, Albert Duruz; my colleagues, Alberto Chirif, Anna Tsing, John Beauclerk, Marcus Colchester, Pierrette Birraux-Ziegler, Oliviero Ratti, Laurent Demierre; nicotinic receptors, Marc Ballivet; dimethyltryptamine information, Olaf Anderson, Novartis; nicotine information, Brigitte Caretti, Switzerland's Federal Office of Public Health; stereograms, Madeleine Siffert, Pascal Siffert; the term "DNA-TV," Kit Miller; child care, Sandrine Arnold, Marianne Santos.

Books: The Cantonal and University Library of Fribourg, Switzerland; The Network of Swiss Libraries; Librophoros, Christophe Piller, Fribourg, Switzerland; Flashback Books, Michael Horowitz, 20 Sunnyside Avenue, Suite A195, Mill Valley, CA 94941.

In Peru: Sally Swenson, Victoria Mendoza, Abelardo Shingari, and the community of Quirishari.

Original fieldwork funded by National Science Foundation (No. BNS

8420651); Wenner-Gren Foundation (No. 4622); Stanford's Center for Research in International Studies.

Specials thanks go to Carlos Perez Shuma, who made an anthropologist of me; the indigenous people of the world, who taught me the most important things I know, who have kept their ancient knowledge despite persecution, genocide and territorial confiscation, and whose ethical standard is an inspiration; my parents, grandparents, and ancestors for the DNA; the global network of life, with a special thought for the plant-teachers.

Permissions and Credits

p. 80. "Anaphase II . . ." From Watson et al. (1987). Copyright © 1987 by James D. Watson. Published by the Benjamin/Cummings Publishing Company.

p. 81. " 'The cosmic serpent, provider of attributes.' " From R. T. R. Clark (1959), p. 52. Reprinted with permission from Thames and Hudson Ltd.

p. 82. " 'Sito, the primordial serpent.' " From R. T. R. Clark (1959), p. 192. Copyright British Museum.

p. 82. " 'Ronín, the two-headed serpent.' " From A. Gebhart-Sayer (1987), p. 42. Reprinted with the author's kind permission.

p. 83. " 'The serpent of the earth becomes celestial . . .' " From C. Jacq (1993), p. 99. Reprinted with the author's kind permission.

p. 84. " 'Here is the dragon that devours its tail.' " From M. Maier (1965, orig. 1618), p. 139. All rights reserved.

p. 84. " 'Ouroboros: bronze disk, Benin art.' " From Chevalier and Gheerbrant (1982), p. 716. Paris, Robert Laffont, all rights reserved.

p. 85. " 'Vishnu and his wife Lakshmi resting on Sesha . . .' " From F. Huxley (1974), pp. 188–89. Reprinted with the kind permission of Aldus Books and the Ferguson Publishing Company.

p. 87. " 'Cosmovision.' " From A. Gebhart-Sayer (1987), p. 26. Reprinted with the author's kind permission.

p. 87. " 'Aspects of Ronín.' " From A. Gebhart-Sayer (1987), p. 34. Reprinted with the author's kind permission.

p. 89. (No title.) From J. Watson (1968), p. 165. London, Weidenfeld and Nicolson, all rights reserved.

p. 92. " 'The DNA double helix represented as a pair of snakes. . .' " From Wills (1991), p. 37. Copyright © 1991 by Christopher Wills. Reprinted by permission of Basic Books, a division of HarperCollins Publishers, Inc.

p. 94. "Liana (*Bauhinia caulotretus*) 'that goes from earth up to heaven.' " From T. Koch-Grünberg (1917), vol. 2, drawing IV. All rights reserved.

p. 96. "The 'sky-ladder' drawing . . ." From A. Gebhart-Sayer (1987). Reprinted with the author's kind permission.

p. 97. " '*Banisteriopsis caapi*, a liana that tends to grow in charming double helices . . .' " From Schultes and Raffauf (1992), p. 26. Reprinted with permission from Synergetic Press, Oracle, Arizona.

p. 102. " 'The cosmic serpent, provider of attributes.' " From R. T. R. Clark (1959), p. 52. Reprinted with permission from Thames and Hudson Ltd.

p. 105. "A magnified section of a leaf . . ." From a photo by Alfred Pasieka. Reprinted with the photographer's permission.

p. 111. " 'Cosmovision.' " From A. Gebhart-Sayer (1987), p. 26. Reprinted with the author's kind permission.

p. 113. "Detail from Pablo Amaringo's painting 'Pregnant by an Anaconda.' " From Luna and Amaringo (1991), p. 111. Reprinted with the authors' kind permission.

INDEX

Gomez, Ruperto, 4–6, 8, 31–33, 109, 154
Gurvich, Alexander, 128–29

Harner, Michael, 53–56, 58–59, 71, 77
Heraclitus, 97
Ho, Mae-Wan, 128
Holmes, Sherlock, 47
Huxley, Francis, 79

Jacob, François, 134
Jakobson, Roman, 135
"Junk" DNA, 100–101, 139

Kekulé, August, 114
Koch-Grünberg, Theodor, 94

Lamarck, Jean-Baptiste, 133
Langaney, André, 143
Lévi-Strauss, Claude, 12, 14–16, 62
Linnaeus, Carl von, 133
LSD, 49–50, 122–24
Luisi, Pier Luigi, 144
Luna, Luis Eduardo, 18, 69

Malinowski, Bronislaw, 12
Maninkari, 24–25, 30, 34, 106–7, 118, 155
Margulis, Lynn, 91
Mayr, Ernst, 142
Métraux, Alfred, 63, 95–96
Monod, Jacques, 134, 138
Mundkur, Balaji, 114

Nash, J. Madeleine, 141
Nicotine, 118–21, 127, 131

Perez Shuma, Carlos, 19–25, 29–35,
    44–45, 63, 96, 107, 112, 118, 121,
    125, 149, 151
Piaget, Jean, 136
Pollack, Robert, 90, 144
Popp, Fritz-Albert, 128
Popper, Karl, 144

Quartz, 64, 128–30

Radio waves, 31, 125
Receptor, 118–19, 123–25, 127, 130, 136,
    140
Reichel-Dolmatoff, Gerardo, 56–57, 68,
    129

Sagan, Dorion, 91
Schultes, Richard E., 10, 41–42, 96
Scott, Alwyn, 136
Serotonin, 49, 60, 123–24
Serpent, 56, 58, 62–68, 71, 77, 79, 81–83,
    85–86, 92–93, 96, 101–2, 108–9,
    112–17, 151, 159
Shapiro, James, 142
Shapiro, Robert, 161
Shingari, Abelardo, 27–29
Snake, 7, 19, 23, 29, 34–35, 50, 56, 59,
    64–65, 69, 79, 81, 83, 92, 111–12,
    114–16, 157
Stereogram, 46, 48
Strassman, Rick, 122
Sullivan, Lawrence, 161

Tangoa, Luis, 82, 86
Television, 4, 72, 109, 116, 124
Tobacco, 6, 20–22, 25, 30–32, 34–35,
    118–21, 125, 127, 129
Townsley, Graham, 61, 98–99
Twins, 59, 62, 66, 71, 98, 106
Twisted language, 98–99, 101
Tylor, Edward, 12

Visual system, 48, 105–6

Wallace, Alfred, 133
Watson, James, 73, 158
Weiss, Gerald, 25–26, 94, 106
Wesson, Robert, 140
Wilbert, Johannes, 121
Wills, Christopher, 92

Yahweh, 66

Zeus, 66–67

BIBLIOGRAPHIC INDEX